D0916961

For my whole family.
But, especially, for my mum.

And the two women stood side by side looking at the slender, flowering tree. Although it was so still it seemed, like the flame of a candle, to stretch up, to point, to quiver in the bright air, to grow taller and taller as they gazed—almost to touch the rim of the round, silver moon.

How long did they stand there? Both, as it were, caught in that circle of unearthly light, understanding each other perfectly.

From "Bliss" by Katherine Mansfield

Part One

Aisling

1

LIKE A DOVE

County Clare, Ireland

NOVEMBER 2012

ORLA AND I walked across the school field. All the other girls in our year were way ahead of us, giggling and jovially swinging their gym bags by the straps. It was that delicate time of the morning. That time when the mist swirls and floats above the ground, whispering itself away and disintegrating into the thin air above it.

"I could've sworn I put them in here," said Orla, rummaging around in her gym bag.

I walked close to her, observing the beauty of autumn. Frozen droplets of dew shimmered on the grass like miniature crystals and refracted the limp sunlight. Burnt orange leaves crunched and mud squelched beneath our school shoes.

With her teeth, Orla removed one of her gray woolen gloves. It flopped out of her mouth like the drooping ear of a bunny.

"I'm telling you, Ash," she mumbled, her teeth clenched around the glove, "Sister Molony is feckin' insane. That assembly this morning. I swear to God."

I stayed quiet and continued to look at her. She just kept

fumbling around in her bag with her bare hand. After a moment, I took the glove from her mouth, fiddling with it and then putting it on. It was still warm from her skin.

"Aisling Delaney, you stolen my glove?"

I nodded and let my lips slide into a broad smile.

"Now we have one each." I showed Orla, wiggling my gloved fingers around in her face. She rolled her eyes.

"There they are. Thank Jesus. Thought I'd left them on the counter."

Out of her bag, she produced two bacon rolls wrapped in silver foil. My stomach rumbled as I saw them. She planted one firmly in the palm of my ungloved hand. It felt like a gush of warm water, thawing the coldness of my skin.

"I dropped by O'Connor's in case your mother sent you in without breakfast."

She was right; I hadn't eaten that morning.

"You got this for me?"

"Yeah, course." She laughed, her sage-green eyes catching me gently.

She turned to study her own bacon roll. Her tongue rested on her glossy rose-colored lips as she peeled back the foil delicately with slender fingers. Then, in one movement, she took a huge bite. It wasn't graceful, but I couldn't take my eyes off her. I watched the white dust collect around her mouth, her jaw clenching as she chewed.

Then I looked at mine. Unwrapping it, I took a sacred mouthful and tipped my head back.

"Fuck me," I breathed.

Orla nodded, smirked, then wiped her mouth with the back of her hand before depositing the greasy remnants onto her school skirt.

"Jesus. Fair play," I said, eating more of it. "This is worth the stitch we'll get for sure."

We were approaching the PE changing rooms by that point, so I had one last nibble and folded the other half of the bacon roll up for later, stuffing it into the side pocket of my gym bag. Orla had already finished hers. Letting out a sigh of satisfaction, she licked the corners of her mouth free of ketchup and crumpled the silver foil up into a ball, rolling it around in her palms.

At the scuffed changing room door, we stopped. I couldn't bring myself to open it. Behind it, all the other girls in our year were getting ready for PE, gaggling and babbling like a bunch of pigeons. They were all far too enthusiastic for our liking.

Orla stood behind me and I turned to face her. On her ski-jump nose, which was peppered with rust-colored freckles, there was still a small dollop of ketchup. I smiled delicately.

"What? What's funny?" Orla muttered.

"Think you got a bit carried away there," I replied.

I wiped the ketchup from her nose with the tip of my finger and licked it off my nail.

We studied each other's faces, our bodies fidgeting. Orla ran her bony hand through the bottom of her copper plait, and then dropped it, linking one of her fingers with one of mine.

I let the moment hang, feeling my skin tingle as she touched it.

"Alright then." My voice was weak. "Cross-country beckons."

I started to twist towards the door, intending to finally push it open, but Orla drew me back with one small tug. I looked at her and something caught in my throat. Our faces were close now. I could feel the warmth of her breath on my lips.

Orla assessed the field around us. No one to be seen.

Leaning in, slowly, like a dove tilting its head, she kissed me. It was as soft as silk on the corner of my mouth. I felt her fingers adjusting my chin so that our lips met completely. They moved together like dancers. I could smell her fresh, wind-brushed skin. I could taste the salty ketchup on her tongue. As we pulled apart, my heartbeat pulsed in the pit of my stomach. Both of our faces cracked into a squirming smile.

"We can't let anyone see us," I whispered.

"I know, Ash," she reassured me. "I know."

2

MUTUAL CONFESSION

THAT WASN'T THE first time it'd happened. Orla and I had kissed once already. It had been about a week before, behind the art building. We'd ended up there after lessons and it had just sort of happened. After that, for the whole week up until we kissed again outside the changing rooms, my palms were sweaty and my stomach a constantly spinning merry-go-round.

During the last few years of school, Orla and I had become much closer. Sure, we'd always been good mates, but something had sort of shifted. I think it was our loneliness that brought us together. We didn't have many friends, but we didn't feel we needed them. We had our books and we had each other.

By the point we had our first kiss, it had been clear to me for a while how Orla felt about me, and, I think, how I felt about her. It was kind of obvious from the way we looked at each other, from how our conversations always had some sort of subtext, or from the way we reacted when our skin brushed up against each other.

Whenever we had a lesson that even touched on homosexuality, Orla would make a thing of muttering to me as we ambled through empty corridors after.

"As if being gay is a sin," she'd whispered to me once as we walked. "How long ago was the Bible written anyway? I mean, catch yourself, it's the twenty-first century. My parents wouldn't even say that sort of thing, and they're Catholic, for God's sake. I just won't have someone tell me I can't be that."

"Be what?" I'd prodded, quietly.

Orla had stared deep into my blue eyes. In that moment, she formed an expression. An expression which said, Isn't it obvious, Aisling? Her eyebrows were raised; her lips were screwed together. It was a look that confessed to me, and my own look succumbed. I confessed straight back.

Then we knew. A mutual confession.

The truth was, I'd known that I was gay ever since I'd seen Niamh O'Donnell climb a tree with bare feet at Aoife McGrath's sixth birthday party and got butterflies. A mundane point of realization, I know, but I try not to overthink it.

To be sure, though, my parents would never be fine about my sexuality in the way Orla's would be. My family wouldn't just struggle with it, and they certainly wouldn't accept it.

• • •

THAT SATURDAY NIGHT, a few days after our second kiss outside the changing rooms, Orla and I had arranged to go to the cinema. It was a sort of date, I guess. Not that I was particularly on board with that kind of thing.

No one would know it was a date, obviously. We wouldn't give anything away in public, just in case someone saw us and ended up nattering to someone else. It's not like they'd care about it themselves; it's just that everyone knew everyone where we lived, and I was worried.

Before I left for the cinema, I was up in my bedroom trying to put on some eyeliner. After school on Friday, I'd caught the bus home, as I always did, and I'd stopped off at the shop to buy it. I'd never worn eyeliner before, or any makeup really, so I didn't know how it worked.

I stared at my reflection in the mirror. The long mop of tangled brown hair, the small nose, blue eyes, dark eyelashes, and my lips, which were apple red from the cold.

I muddled around, trying to figure out whether the eyeliner went on the outside or the inside of my eyelid. Just as I was getting the hang of it, Ma called up to my room asking me to lay the table for dinner. I stopped, having done only one eye. Letting out a frustrated sigh, I slapped the eyeliner pencil down on my bedside table, thinking to myself that I'd do the other one later. The nerves before seeing Orla were like popping candy in my stomach.

I went downstairs and put out the plates, knives, and forks.

"Come help me bring things through once you're done," Ma called from the kitchen.

I finished straightening out the tablecloth. It looked like a big doily. Ivory lace pirouetting around the rim, sprouting stray bits of string where things had snagged it.

As I walked through to the kitchen, Ma spun around and looked me up and down. Her lips were wrinkled from scowling all her life, her eyes the same color as mine. Wafts of spice, wine, and earthy vegetables smoked around her as she moved. She wore her hair in a black bun above big pearl earrings, her charcoal linen apron tied into a perfectly symmetrical bow at the back.

"Sweet and merciful Jesus." Her voice was monotonous as she turned back to the stove. "The eyeliner is a bit much."

Her lips were tight and motionless, as if balancing a tightrope walker. The flames from the hob twitched, hissed, and deflated as she turned off the heat. Regally, she slipped on her oven gloves, one by one, and picked up the big, bubbling casserole dish. It spat and squirted beneath the lid.

"Bring through that bowl of rice and a serving spoon."

Silently I obeyed and took them through to the dining room. It was lit by a brass candle-style chandelier, bespangled with silky spider webs that no one had ever bothered to dust away. Ma had set down the casserole dish and was rearranging the table-cloth, rectifying what she surely saw as my shoddy work.

"What's the occasion?" she asked without glancing up at me.

She started changing the position of the knives and forks.

I put down the bowl of tepid rice, and the white lump of it jiggled like squirming maggots.

"I'm just going to the cinema with people."

"I wasn't aware."

Brushing her hands on her apron, Ma looked at me and raised her eyebrows. I didn't respond, but instead just stared deep into the steaming casserole vapor and sat down.

"Are you calling your father then, or shall I be doing that?"

The three of us ate mostly in silence.

It was always silent, and it was always just the three of us nowadays. My father was a distant, uninvolved man. He had never taken an interest in me.

I also barely ever saw my siblings. At Christmas and Easter, they would come home for a while, but otherwise, they mostly stayed away.

I was the youngest of four. Sean was the eldest, eight years older than me; then there was Jack, six years older, and Mary, who had left home for college when I was fourteen, abandoning me completely. The last thread attaching me to something outside of this nightmare had snapped when Mary left.

All three of my siblings lived and worked in Dublin now, only a few hours away on the other side of the country. I never spoke to them, and I don't think Ma or Pa did much either. We had never been close given everything that happened to me growing up. They had just ignored it, blanked it out, just like my father.

I sat there at the dinner table and pushed the casserole around my plate, then asked if I could be excused.

"But you're not finished." My mother dabbed at the corners of her mouth with a cream-colored napkin.

"You haven't eaten much. It's a waste," murmured my father.

"I'm not hungry, and I'll be late if I don't get going."

"Clear your plate first," instructed Ma. "And what time is it you'll be back?"

"I dunno, around eleven?"

An unimpressed nod. Ma sat upright, decorous, holding her cutlery, elbows pointed out. The piece of beef on her plate was being sawed, the toughness of it resisting incision.

I took my plate through to the kitchen and scraped the stodgy, conglutinated lump of food into the bin. The heavy bulk sank to the bottom of the bag. I watched it fade away, then put my plate in the dishwasher.

Walking up the road to the bus stop, I tucked my hands deep into the pockets of my long black coat. Breath fogged out of my mouth, and fat drops of rain plonked down from the sky, hitting me like a smacking hand.

I arrived late to the cinema and brushed off the drizzle that had gathered on my dark hair. In the foyer, I saw Orla up at the ticket stand. The flashing lights, red and white and candied neon colors, blurred around her perfectly still figure. Like fire, her hair set the grayness of the place ablaze. She caught sight of me, and that crinkle on her nose emerged as her lips turned upwards. I couldn't look away from her. I was in her grasp, and she knew it.

"Sweet or salty?" she mouthed from across the room.

I watched her lips moving like a paintbrush on a canvas, and I decided that perhaps I was on board with this being a date after all.

• • •

ORLA SWUNG HER car into the curve of my road. After the film, we'd driven along the country lanes, humming along to the late-night radio music. When we got close to my house, I reached over and turned down the volume, causing the soft, crackling voices to mellow.

"Shall I pull in at the end there or go next to your house?"

Everyone should be asleep, I thought to myself.

"Outside the house should be alright."

The car pulled up, and she let the engine rumble for a second before switching it off. The windows to the place looked impossibly dark and full of shadows. No glint radiated from inside those walls. I watched it for a moment, the place that was meant to be my home. I didn't want to go inside.

"That was fun, Ash, even if the film was a bit shite."

"Yeah, yeah, it was great craic, and cheers for the lift back."

I turned away from the darkness and peered into her bright eyes. They were the color of seaweed gliding on curling waves in the full moonlight.

I leant over to hug her. As I held her, I smelled her citrus hair, and as I drew back, our noses almost touched. Orla stroked my pale face with the back of her cold fingers; then she kissed me. Her lips touched mine, gently, then more decisively. Her hands moved to the back of my neck, her thumb stroking below my earlobe.

"Jesus," I breathed as we pulled apart.

"I know," Orla almost giggled, breathlessly.

"But what if they"—I dipped my head towards the house, cleared my throat—"found out?"

Her eyes widened, owllike. She put her hand on my knee and rubbed the inside of my thigh. Both our breathing was heavy, and I could feel the air between us vibrating.

There was complete silence. It floated like an autumn leaf swaying slowly in the breeze, down from a tree to the ground.

Orla knew about my family. Not the details, but roughly speaking. She'd guessed it when we were much younger. At the time, I hadn't denied it. I'd just told her never to ask me about it again, and she hadn't.

"You look really beautiful, Ash."

That was all she said to me.

I pushed my lips together and shut my eyes momentarily as I breathed out; then I said good night, got out the car, and walked up the driveway. As I entered the house and clicked the door shut, I could smell the normal, rancid waft. Incense, fried onion, dusty carpets. I crept into the hush.

It was then that I heard it. The sound of a glass being put down on a table. The clink of ice. A swallow.

"Aisling."

The low, grassy tones of my mother's voice croaked like a ribbiting frog. My heart ate up my chest and crawled into my throat.

Beating, beating, beating.

I poked my head around the corner of the sitting room to see her lying there, strung out on the sofa. There was a single strip of light across her face from where the curtain let in slim rays of moonlight. I could see her raw, bloodshot eyes. Her mascara was painted like ripe, plum-colored bruises under her eyelids. The smell of gin in the air, soaking into the carpet, reeking out like a chorus of shrieking gulls.

This, to me, was the real her. The woman behind the demure, pious exterior. It always had been, and it always would be.

"I saw you. In that car out . . . over there. I saw," she said, pointing in the general direction of the road. As she picked up her glass, the pings of the ice clanged like church bells.

"What you doing, kissing a girl?"

Her head tilted, wolflike as she spat out the words. Moments passed and just the sound of her drunken breathing hung in the room. Her foot tapped the sofa cushion. Flints of sparkling dust drifted upwards, tiny bubbles bursting.

"Come sit."

I shut my eyes, firmly, and shook my head.

"Fanny-sucking bitch," she choked. "Sit."

I felt that her face was right up close to me, but I didn't know where it was. My eyes were so tightly shut.

Draining the last few drops of her drink, she thwacked the glass down on the table. As if it were mouthwash, she rinsed the cold, sharp alcohol around her gob and swallowed.

For minutes, nothing was said. I just waited for it to happen. But then, after a while, I heard her starting to cry. I could almost feel the tears trickling down her skin. I was silent.

Silence had always been commanded, but it had also become a coping mechanism for me. Movement and speech only ever seemed to make things worse; it gave her the sense I was fighting back.

My mother looked up at me through the darkness. Beneath her tears, I thought I could see a hint of her eyes confessing something to me, too.

3

MASS

"AISLING!" MA YELLED from downstairs. "FIFTEEN MINUTES!"

As I came around the next morning, I rolled onto my back. My wooden bed creaked as I stared at the mud-colored curtains on the other side of the room. Pinprick holes in them let in streaks of dull sunlight.

The draught which blew in under my door carried the scent of coffee and burnt toast, and I could hear classical music, Mozart, humming in the background. I peered up at the ceiling, where minty green glow-in-the-dark stars had been stuck since I was a child.

My mind wandered. I thought about last night, trying to erase the part with Mother from my mind. It was always like this the following morning; everyone, me especially, trying to pretend as if nothing had ever happened.

I thought back to before that, back to the car, back to Orla. Orla's hand on my leg, her lips on my lips, her thumb pressing the skin behind my ear. I shut my eyes and put my hand down under my knickers. I let my body weaken at the thought of her.

"Aisling, for Jesus's sake, get yourself up!"

Ma opened the door to my room harshly.

I scrambled, jerking into an upright position.

"Shit."

Our blue eyes locked, and she held the door handle so tight that her knuckles were like freshwater pearls.

"Ten minutes," she spat.

Eyes narrow, lips tucked together, she penetrated me with her gaze. The door slammed behind her. I listened to the sound of her marching down the stairs.

I swung my legs out of bed, my chest tightening and going hot from the remnants of her look. My breath pulsing, I went to the mirror and pushed my dark brown hair flat at the top of my head, then combed it over my shoulders. I realized, as I stood there, that I'd never put eyeliner on the other eye. Orla hadn't pointed this out the night before, so's not to embarrass me. Momentarily, I buried my reddening face into my sweaty palms.

I felt sick to my stomach, but I had to keep going.

Face out of palms. Smudge away eyeliner residue. Put on Sunday dress and cardigan. Find shoes.

I went down to the kitchen. Ma was standing at the sink wearing yellow rubber gloves. She was scrubbing a pot in steaming-hot water, even though it already looked clean. Her eyes were bloodshot and dry. I could smell the stench of her alcoholic breath and the attempts to cover it up with toothpaste and mouthwash. My father was sat at the table reading the newspaper, sipping at a full glass of orange juice.

"There's a loaf on the table if you want," he mumbled.

Ma didn't turn. She just continued to scrub.

I cut a slice from the loaf. The top of it looked shiny, like a shimmering ice rink. On the inside, the currants glimmered like petite amber and garnet stones, perfectly polished. I put the bread in the toaster and smelled the sweet fruits as they started to melt and singe. Pa looked up at me over his newspaper, his glasses precariously poised on the end of his long nose.

"That dress is a bit short, is it not?"

"I wear it every week, it's the only one I've got that's smart."

Ma still hadn't looked up from the pot she was scrubbing.

"Fiona, will you take a look at this now?"

Ma kept scrubbing, speeding up her arm movements as if she were accompanying the upbeat part of Mozart's Violin Concerto no. 3.

"Fiona?"

My father put down his newspaper and moved his chair back. It scraped on the floor, making a noise like fingernails on a blackboard.

"Go change," Ma muttered quietly, rolling the pot around in a stream of piping-hot water. The steam condensed onto the glass cabinets above.

"But—"

CRASH.

The heavy metal pot dropped into the sink onto a bowl.

"I SAID, GO CHANGE, AISLING," she screamed, turning, her hands shaking. Her cheeks shone like blushing apples.

"PUT SOMETHING MODEST ON AND BE IN THE CAR IN TWO MINUTES."

Ma wiped her forehead, the wet glove leaving a streak of water on her clammy brow. Plucking at each finger of the gloves, she flayed them off her shuddering hands like snakeskin.

As silence filled the room, the bread popped up in the

toaster. It was black and burnt. Smoke hissing from the singed currants. I looked at it and yanked it out. I had a nauseating feeling at the bottom of my stomach, and I didn't feel hungry any longer.

As I opened the bin to throw the charred piece of toast away, I saw a broken crystal glass perched on top of the other rubbish. I thought about my father coming downstairs that morning, sweeping up the glass with a dustpan and brush, mopping up the spilled alcohol, and chucking it all away. It was to be ignored, forgotten, and thrown out of sight.

• • •

A FEW MINUTES later, I got into our maroon Citroën Synergie wearing black trousers, a shirt, and a blazer which Mary had handed down to me a couple of years before. My parents were already sat there, waiting.

We drove to church in silence.

There was a small queue for the car park when we arrived.

"We're late. Why don't you two get out. I'll park."

"See you in there," said Ma, unplugging her seat belt and getting out the car.

I followed, but even by the time I was out, Ma had walked way ahead of me. I moved swiftly to catch up.

"Ma."

"What?"

"You haven't said anything to Pa, have you?"

My breath was uneasy as I jogged to keep up with her. Around the corner we went onto the pathway, past the lofty, naked trees, the wonky graves, and up towards the big wooden archway that led into the church.

"No." Her word cut down like a guillotine.

Ma and I slotted into one of the pews, right at the back where no one else could see us.

"Welcome to this morning's Mass," belted the priest with open arms.

As he continued his greetings, Ma leaned over and lifted my knotty, sooty hair above my ear to whisper something. To others, it must have looked like something so trivial. Like she was telling me what we were having for dinner, or what number the hymn was, or what she thought of the flower arrangements in the church. Instead, she slowly let her lips curl around the words.

"You deserve this."

She dropped my hair back over my ear and retracted her callous fingers.

I knew something was coming. I'd known it last night. This sort of thing had happened most of my life, and I could always feel it looming. And that fear had, over all these years, turned into a different sort of sensation. It felt, to me, like when you're just about to vomit. Wanting it to happen, almost, just so that it'll be over.

The hymn started.

Let all mortal flesh keep silence,
And with fear and trembling stand.

Pa joined us in the pews, sitting next to my mother. Even if he noticed whatever she was about to do, he wouldn't intervene. She knew that. I knew that.

For with blessing in his hand.

My mother grabbed my hand, tight. She squeezed it so hard I felt like my bones were being crushed. My breathing stopped. I knew, after all that time, to stay quiet. To suck up all the pain into myself, to keep silent or else there would be consequences.

Lord of lords, in human vesture,
in the body and the blood.

Her fingernail pierced my skin. Blood. She clasped the back of my hand, and her thumbnail dug into my palm. The nail broke the skin and went in, deeper, deeper. My breathing got heavier. I wanted to scream, to cry out. But I didn't. I couldn't scream. I couldn't turn. I soaked up the agony like a sponge. I shut my eyes tight as I felt the sting of her razor-sharp nail digging into the palm of my hand. The piercing getting deeper, deeper, deeper.

Cherubim with sleepless eye,
Veil their faces to the presence,
As with ceaseless voice they cry.

I could feel the warm blood on the skin of my hand. The sting getting worse. The nail going further and further into my palm. She twisted it around. My body began to shake, convulse; tears welled in my eyes. And as I kept them shut, I imagined her face. Her tongue, positioned at the corner of her mouth, salivating at the thought of my hurt. Her breathing deep. Her eyes wide and blazing, reflecting the flames of hell. Her hair, springing out of her neatly combed bun, as if she'd been electrocuted. This was the face of my nightmares.

Alleluia, Alleluia,
Alleluia, Lord Most High.

As the hymn finished, she turned to me.

“There you go, my cherubim.”

Her voice was like a whisper carried only by the wind.

I turned. My face twitched. I could feel the tears on my cheeks, rolling down and away like stones.

Licking her thumbnail, my mother swallowed my blood whilst it was still warm, then faced back towards the altar.

4

WEEPING STATUE

JULY 2013

EIGHT MONTHS LATER, I sat on a small wooden chair in the courtyard behind my parents' house, reading a book of poetry. On either side of me there was an array of sad potted plants that needed watering and, behind me, a washing line strung from fence to fence. There were various T-shirts, trousers, and dresses of mine hung up with wooden pegs. A full cup of coffee was perched on the ground by my feet. It steamed and floated upwards, producing ribbons, which swirled in the summer air.

It was around noon and the courtyard was swamped with sunlight. The day was hot for Ireland. I wore a dark linen dress with a V-neck so that my pale chest could get some sun. My hair tumbled down my back in a long plait. I could already feel my cheeks getting hot, so I'd put on an old cap which used to belong to one of my brothers—I couldn't remember which.

I turned the page of the book slowly, then picked up the mug and swallowed a short sip of the coffee, letting the bitter caffeine hit the back of my throat before placing it back on the ground.

It was the summer holidays, and it was a Thursday. Thursdays were my favorite days because I had the house to myself. My father was at work, as he was every weekday, and my mother worked on Tuesdays and Thursdays at the local dentist as a receptionist. Given that I usually had a shift on Tuesdays at a store in town, a part-time job I'd taken just for the summer, Thursday was the only day I had alone. I could relax, read, do my laundry, and sometimes even try to write some very bad poems. These dire attempts were all hidden away in the back of a notebook which was jammed behind a painting which hung in my room.

As I sat there, I heard a buzzing noise. It was my phone vibrating, but no one ever called me.

I looked around for it and saw my phone on the ground near the coffee mug. Looking at the name there, it read "Joseph McMahon." Last autumn, I'd changed Orla's name on my phone to Joseph McMahon when we'd first started dating, just in case someone saw my screen and got suspicious. Joseph was this real annoying lad who Orla and I had been to primary school with, so we'd found it hilarious at the time.

I hadn't spoken to Orla since she'd ended things with me in May, right before our final school exams had started. We'd been together for a few months by that point, but I knew she was finding all the sneaking around too difficult. It was the driving out to remote places in her car to be together, not being able to speak on the phone, having to hide our relationship at school, needing to conceal our affections for each other in public. It had all become too much for her and she didn't want it anymore.

"You can come to mine," she used to say. "My parents would be fine with it if I spoke to them."

"No," I used to respond. "What if your parents tell my parents? Or if word got out? You know how things get around here."

She'd bitten her lip as I spoke.

"No one would care, Aisling. Plus, I feel like we're doing something awful," she'd mutter, "and we're just not."

But I'd said no, no, no. No, we couldn't do that. No, I didn't have a choice. No, I had never had a choice. If my mother got a whiff of anything, I knew something worse than her nail in my hand at church would happen to me, and Orla knew it, too.

Eventually, she told me she couldn't be with me because she cared about me too much; she couldn't handle the responsibility, the pressure, the required secrecy of the relationship.

She'd seen my bandaged hand after church back in November, catching a glimpse of it at school. Holding my hand in hers so my palm faced the sky, she peered into my eyes, waiting for an explanation. I just shook my head and told her not to ask.

I was devastated when things between us had ended. I'd told her it would be different when we went away to university. It would be a new life. We wouldn't have to sneak around; she could visit me without anyone knowing. But deep down, I knew it was right for things to end, and after all, perhaps it was better for her not to be with me.

Exams had come and gone now, and we were waiting on our results. Hopefully, I'd go to university in Edinburgh, she'd head off to Cork, and we could both start afresh.

I stared at the name on the phone screen. I felt the buzzing in my palm, and then, quickly, I walked inside the house so none of the neighbors could hear. They probably weren't even listening, but it was for my own peace of mind.

I answered the call.

"Hello?" My voice shook.

"Ash, er, it's me," Orla mumbled, her voice timid and small.

"Yeah, I saw."

"Right," she said. "Is it an OK time?"

"Sure." My throat had gone completely dry.

"Sorry," she said. "I shouldn't have called. I know it's weird. I just, I have some of your stuff is all, things you left in my car or whatever. I'm heading off on holiday tomorrow for a while and then we'll be going our separate ways when I get back, so I thought—"

"Alright, well, now is good."

"Now? What, drop it to yours, now?" Orla asked me, a hint of shock in her voice.

"Yep, I have an empty place, if that works?"

• • •

A WHILE LATER, after I'd frantically brushed my teeth and tidied the house, I heard a knock at the front door and went to open it.

There was Orla. Beautiful wide green eyes, freckled nose, ivory teeth glinting through the crack in her lopsided lips. She wore a Gap T-shirt, trousers, and in her hands she held a small brown paper bag. Her hair was in a long, copper ponytail, placed like satin over her shoulder.

I wanted to collapse into her, to tell her to take me away, to tell her I was sorry, to reassure her that it would all get better.

But I just smiled, pressing my lips together.

"Nice cap." She grinned.

"Cheers."

I took it off and placed it on the sideboard.

"I brought your stuff," she said, holding it out robotically.

"Ah, cheers."

I took the bag, opened it, and examined the contents. There wasn't much in there. Some pants, an old CD, a crocodile hair clip. Things I probably wouldn't have missed.

I cleared my throat and looked back up at her, waiting. She lingered there on the doorstep, rocking back and forth on her feet.

"Do you want a coffee or—?"

Dimples pressed into her cheeks.

"I'd love one."

I stepped back so she could come into the house, and then I shut the door behind her. When I turned around, she was poking her head in through various doors. The sitting room, the dining room. This must have been the first time she'd been to this house since we were very young. I knew for sure that she'd been here for my eighth birthday party because there were a bunch of photos from the occasion. Among the plates of jelly, the balloons, the party dresses, my eye always went to Orla in those pictures. I remember she wore a kind of pea green party dress with polka dots on it and a shining gold sash.

"I haven't been here in years, but it's exactly how I remember it."

I put the bag down on the sideboard, on top of the cap, and walked past her through to the kitchen.

"It's the same really, except it's a bit more scuffed up. Like all the marks on the walls."

"Oh yeah?" she said, coming in through the kitchen door and scanning the space.

"What about this one?" she asked, pointing at the noticeable dent on the wall opposite the back door.

I went over to look at it and laughed awkwardly. "Oh, Jesus." I cringed.

I moved over to the kettle, filled it up, put it on the base, and flicked down the switch.

"I had a wobbly tooth when I was wee, and Mary told me that

if I tied string to the tooth on one end and the back door on the other, she'd slam the door shut and the tooth would come out. She goes and does it, slams the door so hard that it not only rips out the tooth and damages the gum"—I ran my tongue over the spot in my mouth—"but I fell back so hard into the wall that it made that mark."

Orla smiled and let out a breath of half-hearted laughter.

The kettle came to the boil, then clicked off. We stood there in silence facing each other, our hands resting on the countertops behind our backs. She tapped her small fingers there and sniffed.

I turned abruptly and got two mugs from the hooks on the dresser.

"Grab the milk from the fridge, would you?"

I spooned some coffee into the cafetière and poured the boiled water over it.

Orla fetched the milk and came to stand next to me. As she decanted it into the mugs, I turned to look at her profile. She put down the bottle on the countertop and turned towards me, too. We watched each other as if no time had passed at all. It was as if the few months we'd been apart had dissolved. She rested her forehead on my forehead, just like she used to do. I could feel the breath from her nose on my top lip. I could feel my chin going weak, my eyes filling, my vision blurring. I swallowed. Orla pressed her lips into mine.

"How long are your parents gone?"

"A few more hours. Why?"

She linked our fingers and led me slowly upstairs.

Seeing her in my room, she seemed so out of place. Carefully, she removed her clothes and then mine, kissing my body lightly as she did, looking at each part of my skin, running her fingers over my scars. She caressed my neck, shoulders, cheeks, ears. She gently guided me down to lie on the bed with her and then rolled me over, so I was underneath her.

Then a click, a twist of a lock. The front door opening, closing.

We froze. Our eyes hooked.

"Hello?"

A voice downstairs. I couldn't tell whose. Our breathing was in sync, our stomachs rising, pressing into each other, then falling. I could feel the wet from her saliva still on my body. I put my forefinger to my mouth, indicating for her to be silent. Then I mouthed at her to get into the cupboard as quietly as she could. I barely had to form the words with my lips; she understood. Orla's eyes went glassy as she nodded.

I put on my clothes as the voice came again.

"Aisling? You here?"

"Coming," I shouted. "Just coming."

I watched Orla put on her top and trousers, then get into my cupboard. I felt queasy and light-headed. But I had to be normal. I had to act normal.

I checked in the mirror, unruffled my hair, and tucked it behind my ears. Rosy-cheeked and with a film of sweat on my skin, I walked out of my room and down the stairs as if in a trance. My vision was hazy, and my breath seemed to jolt on every step. My heart was beating so hard it was all I could hear.

There was the body of the voice, right by the front door.

"Sean?"

Sean had placed a small suitcase down and was assessing the sideboard.

"That's my old cap. Why's that there?" he said, puzzled, pointing at it. "And what's this?"

He picked up the brown paper bag which was sat on top of

the cap. I hastened towards him and snatched it away abruptly. Sean held his hands up in surrender.

"Alright! Grand to see you, too, Aisling." He grimaced.

I hadn't seen Sean since Easter and we hadn't made any effort to speak since then.

"How come you're home?" I asked.

"Got Connor's wedding tomorrow in town, didn't Mam tell you?"

I shut my eyes for a moment and pressed the bag to my chest with clammy hands. "No, she didn't," I responded.

"Just here tonight and tomorrow and then I'll be away."

"Right."

I gazed at him in silence and complete disbelief. Of all the times he could have arrived home. I was beginning to feel even more dizzy, like I might faint any second.

There was a noise from upstairs. I shut my eyes again. It was the fall of a hanger in the cupboard or the bash of a head on the wall. No, no, no, I thought to myself. No, this isn't happening.

"Are the others home? Ma said it'd just be you around, but . . ."

I didn't move, didn't even open my eyes.

"Aisling, is someone here?"

I shook my head this time and, slowly, I let my eyelids curl upwards. "No," I murmured.

We stood there, awkwardly. I could feel my whole body contracting, my vision melting into wax and light. Two pins pressing into my temples.

"So, how're you keeping? Spoken to Mary or Jack recently?" Sean asked me.

"No, I haven't done. Have you?"

Sean turned his lips downwards and shook his head. In the pause, he narrowed his eyes.

"Aisling, have you got someone upstairs there?"

I pressed the brown paper bag further into my chest to try and hide my pounding heart, which felt like a hand punching through my skin.

If Sean had come home and Orla had been here, just having coffee or something, he wouldn't have thought anything of it. Just a mate, he'd think, she's obviously just a mate. It was the fact I'd tried to hide it which made it so obvious. And anyway, even if he didn't suspect, I didn't want him blabbering to Ma. She'd figure everything out; she always did.

Just as I was deciding what to say, what to do, before I could stop him, he was shaking his head, muttering to himself, and marching up the stairs towards my room.

"Sean," I protested.

He peered around my door frame and then glanced back at me before entering my room. I'd just reached the top of the stairs. There was the sound of breathing, and he knew where it was coming from. He moved towards the cupboard and slowly opened the door.

Orla toppled out, her face contorted, her eyes damp from tears. Her clothes were on, but her T-shirt was backwards and her trousers weren't done up. I looked at her helplessly, trying to speak with my eyes, desperately trying to tell her the things I wanted her to understand.

She was beautifully carved. Perfect, like a weeping statue. I could feel her marble-like smoothness slipping from my hands, slipping away from me. She couldn't do this again; I could feel it.

Sean glanced at my unmade bed and then turned to me. He knew what was going on.

"What the fuck, Aisling?" he raised his voice at me.

"Um, uh—"

"Right," Sean said, turning back to Orla, who was stood completely still.

"Please, Sean," I begged.

Orla scurried past me. Before I could take her in again, she had disappeared. I listened to her footsteps down the stairs, through the corridor, her hand opening the door, the slam. I readjusted my sight. It was misty; I couldn't see much really. I wanted to cry, but I felt numb and completely empty.

"Please, please don't. Please don't tell."

I could sense the tears on my cheeks, but I didn't know how they were there. I didn't recall conjuring them up or letting them loose.

Sean stared at me, his nostrils flared, his eyebrows ruffled together. It felt like hours on end before he uttered another word, as if we were caught in some ethereal space together.

"I won't. But we will never speak of this again. You hear? Never." His words stuck in to me like pins.

I bit my teeth into my bottom lip and felt my head quivering. I tried to stop myself from crying even more, but I couldn't. As Sean exited the room, I felt myself shut the door and fall face-down on my bed.

I didn't hear from Orla again that summer, and I was sure I'd never see her again.

Part Two

Maya

5

INSECT

London, England
JULY 2013

I STOOD AT the Northern line platform in King's Cross tube station. Baggy black clothes draped over my body. I wore bright red lipstick, heeled boots, big hoop earrings, and carried a small black handbag. I'd tied my hair back into a slick ponytail after straightening out all my curls. I liked it better that way; it made me feel sophisticated.

It was a Friday night around half nine, and the station buzzed with people. Loud music and drunken shouting bounced off the curving walls, but I felt that my stillness and outward confidence created an untouchable bubble.

I got my phone out my bag and unlocked it, clicking on a text from Ethan that I'd already read six times.

> Naomi tells me you're on a date. Not jealous I promise. Come over if it ends early—Naomi and Tim are at mine until late x

I smiled and sighed, locking my phone and putting it back in my bag. I had just been on said date, and it hadn't gone well, so I was going to join them.

I hadn't really wanted to go on a date, but I was sick of everyone asking me if I was seeing anyone. For some reason, people were constantly prying about what was happening in my love life and I had nothing to tell them.

In theory, I had everything I could possibly need and want. I had lovely parents, wonderful friends; I'd just finished my first year of university at Edinburgh and was loving it, and I was having a great summer back with all my mates from home. But there was something missing, and I wasn't really sure what it was. I felt guilty about this feeling, like I wasn't appreciating what I had. But for some reason, I couldn't shake it off, no matter how many times I told myself I should be grateful.

I'd said all this to Naomi, a friend of mine who I'd known since nursery. After deciding for me what might be missing, she had set me up on a date with some guy who knew her brother. He was called Fred. He had sounded quite nice. She told me he was handsome, clever, a bit older than me. He'd just graduated actually and worked in finance or something like that. The fact he worked in finance or something like that made sense when he suggested we meet at one of the most expensive sushi restaurants in London. I was a vegetarian who didn't eat fish and a student who didn't have much money, so sushi wasn't ideal, but I had agreed anyway because I couldn't be bothered to suggest anything else.

On our date, I discovered that Fred didn't think my degree in English and Scottish literature would be lucrative, and that he didn't like books of any kind. This didn't set us off on the best foot. He then told me that poetry was a waste of time. Another slight hurdle, given that poetry was one of the only things I really cared about. Instead, Fred spoke to me about his university days at Durham, his ex-girlfriend, and how much money he was making. I handled all this by drinking a lot of alcohol and leaving the restaurant at the first polite opportunity.

My lack of love life wasn't down to not fancying anyone. I'd had the odd crush on boys at school and even a few at Edinburgh, but it just never seemed to go anywhere. I knew that, more likely, my lack of love life was down to the fact that there was one person who I'd always liked and who took up most of my focus.

Ethan, Naomi, Tim, and I had been mates since primary school. We used to run around the playground together, finding different types of bugs and performing marriage ceremonies between them. Ethan and Naomi had ended up going to private school as we got older, and Tim and I had gone to the same state school in West London. Miraculously, though, we'd all stayed friends somehow. Close friends.

As we'd grown up, Ethan had become more and more handsome, funny, charming, and every time I saw him, I became increasingly nervous. Since we'd been about sixteen, I'd seen him in a completely different light, and my feelings for him were only becoming more unbearable and more painfully obvious. Nothing had ever happened, though, and I was sure it never would.

At the pub the week before, I had spent minutes on end simply staring at Ethan across the table. He had been laughing, just laughing. Dimples pressing into his lightly stubbled cheeks. Naomi had been talking to me as I sat there, staring, and had clicked her fingers in front of my face.

"Maya, I've asked you if you'd like me to set you up with someone like twice now, and you're just staring into the distance."

Naomi's eyes had gone so wide that her mascara smudged just below her eyebrows.

"Alright then." I'd shrugged, giving a pathetic smile, spinning my wineglass between my darkly painted nails.

Naomi had slurped at the melting ice cubes at the bottom of her glass, her eyes still glued wide open.

"OK, OK, I've got you. I'll give Fred your number."

Ethan was entirely out of my reach. I knew that. But, having said that, I had sensed things changing that summer while we'd both been home from university. I caught him glancing at me a lot. I could feel him coming to stand next to me at parties. I noticed he was texting me more often to see what my plans were or coming up with excuses to see me. But perhaps I'd been imagining it.

The rush of smoggy wind passed through my hair and clothes as the tube rushed through the tunnel towards the platform. The rattle of its slowing resounded in my chest.

I got into the carriage and sat down, feeling a bit drunk. I hadn't eaten much sushi, as predicted, but had drunk a lot of wine, also as predicted. I reapplied my lipstick using my phone screen as a mirror, then wiped below my eyes to get rid of any mascara smudges.

The doors shut and the tube started to move.

I felt my tummy twisting.

Ever since the end of school, a sort of munching feeling constantly ate up my stomach and my muscles. It arrived about five minutes after I woke up, dipping down during the day, and then rising again in the afternoon to hiss around my body like an annoying insect nibbling away. It made me feel liquid and cloudy, but I'd learnt to hide it well.

I observed the people around me, trying to distract myself. A man on the way back from work, his briefcase in hand and newspaper tucked under his arm, face gloomily staring into the distance. A woman wearing bright pink lipstick and sunglasses, her perfectly groomed chihuahua sat on her lap. A giggling group of girls with flared jeans and cartilage piercings, their water bottles obviously full of gin and mixer.

I turned to look at the other end of the carriage. A girl about my age was sat there, reading her book. She wore white linen trousers, a floaty turquoise top, and a headband. I watched her. The flimsy light above her head flickered, hiding, then revealing the freckles at the top of her nose. I looked at how she turned the pages of the book by curling up the paper into a perfect arc, and then brushing it over and locking her sight onto the next page of text. Her deep brown eyes followed the lines of the words like grass fluttering in a gentle breeze.

The tube slowed down, the screech of the brakes squeaking through the train, piercing my eardrums.

"This station is Camden Town, Edgware branch. Change here for all stations to High Barnet and Mill Hill East from platform three. This train terminates at Edgware."

The doors opened. The girl shoved her bookmark in between the pages, stood up, and got off. The doors shut.

"The next station is Chalk Farm."

I stared into the space where she used to be. We kept moving.

"This station is Belsize Park."

The train slowed again and came to a halt. I shook my head to get myself out of my drunken trance, made sure I had all my stuff, and got off. As I entered the ticket lobby, a text came through from Ethan.

Glad you're joining. You almost here? X

I scanned my Oyster card, slipped through the gate, and messaged back.

I'll be five, just walking from the station x

I tucked my phone into my bag and crossed my arms.

People flooded through the gum-speckled streets of London. I popped across the road to buy a bottle of wine from the off-license, which I then tucked under my arm as I walked. I strode up the hill, passing by the cafés and restaurants in Belsize Park. People were sat with their frosty beers in the softness of the summer evening, lit up by the orange glow of patio heaters. Laughter and shouting popped and fizzed in the air. I got to the old church and turned up onto Ethan's road. When I got to the house, I walked through the small metal gate, down the paved path, and then climbed the palatial stone stairs to the big black door. I knocked, and after a couple of minutes, he answered.

A cigarette was tucked behind his ear, a half-full beer bottle in his hand. I could smell the mixture of sharp booze, barbecue smoke, and bitter deodorant.

"Bad date? What a shame." His dimples pressed into his cheeks like soft furrows on a field.

"You're terrible."

He held out his arm and guided me into the house. As I passed by him, I handed him the bottle of wine. He thanked me and closed the door.

"But really, how bad?"

"Oh, it was great, yeah. He told me that my English degree was pointless, that I was a fussy eater, and rounded it off by telling me that he was still in love with his ex-girlfriend."

"Naomi did a good job setting you up then."

"I think she should do it professionally."

I hung my coat up and then my bag, taking my phone out before turning around to face him. I sighed through the jitters of being near Ethan, straightening out my clothes and running my hands through my sleek ponytail. Our eyes met.

"Wine?" he sighed.

"Dear God, please."

Ethan led us through the corridor. The floors were tessellated with diamond shapes in various shades of gray, and cream radiator covers were topped with expensive candles and clay pots. He pushed the door to the kitchen open. It was cool but warmly lit, the cream and marble surfaces stretching out below the skylights to the glass doors which opened out onto the patio. There were glints of light from the outdoor lamps and the tips of cigarettes being sucked. Naomi's and Tim's laughter reverberated into the house.

Ethan got a glass out of one of the cabinets, then took a crisp bottle of white wine from the fridge, replacing it with the tepid bottle I'd given to him.

"Your parents away?"

"France." He smiled, pouring the wine. "I'm going out Sunday morning."

He passed me the glass and held on to it for a moment too long before letting go.

"Oi, you two," screeched Naomi from outside. "Stop flirting and get out here."

I laughed awkwardly, feeling my cheeks go warm.

The evening ticked over as it always did. We drank and reminisced about the good old days of school, talking about tragic sports matches or nights out where one of us had done something horribly embarrassing.

During a break in the conversation, just as our laughter dissipated into the late-night air, Naomi blew a line of cigarette smoke up into the moody sky and then peered over at me apologetically.

"Sorry about my setting-up skills, Maya. Was it truly awful?"

"I'm plotting my revenge."

Naomi flapped her hands and bounced up and down in her seat. "No! I'm sorry. I normally have such a good instinct for these things."

"Anyone want more to drink?" interrupted Ethan.

"Last week Ethan went out with my friend from Brighton and that was good. You gotta admit. You know her, Maya, actually, you met her when we were out last week, remember. Rose? She's got the, like"—Naomi paused, took a drag of her cigarette, and mimed short hair with one hand—"short brown hair and those, like, big blue eyes?" Naomi blew a trail of smoke out of her lips.

"Oh, yeah." My head felt like a sponge filling with water.

"Mate, you went out with Rose?" said Tim, grabbing Ethan's shoulder and shaking him. Ethan's body stayed slack.

"Does anyone want more to drink then?" I said, standing up quickly.

I went inside without listening to their answers and put my empty glass down on the counter. After standing there for a second, I walked to the toilet under the stairs. I didn't need the loo, but I locked myself in and sat on the toilet lid without pulling down my trousers. I slumped my head in my hands. I felt so stupid for having reapplied my lipstick on the tube, for coming here, for thinking there was any possibility of something happening. I'd been such an idiot.

When I walked back into the kitchen five minutes later, Ethan was at the sink, washing up a glass.

"You OK there?" He turned off the tap and dried his hands on a tea towel.

"Yeah, sorry, I was just—"

"That stuff with Rose, I mean, she's great, but it was just a bit of fun. I wanted to say that. It didn't really mean anything to me or to her, I don't think."

"You don't need to explain," I laughed. "It's fine, we're just mates, and anyway I'm just . . ."

As I spoke, he walked towards me, getting so close that it stopped me from talking. Slowly he leant in and kissed me. My head was gently pushed backwards, but he held my neck softly and ran his fingers across my skin. Kissing him felt like resting my head on a pillow, almost as if we'd done it before. He pulled back.

"Sorry," he whispered. "You were saying?"

I opened my eyes. I knew that face so well, but I'd never been this close to it.

"I should go."

"Maya."

I walked into the corridor with the tessellated tiles and grabbed my small bag from where it hung. I put on my shoes and snatched my coat from the hook.

"Maya, I'm sorry." He followed me.

"That's OK, honestly, I should probably just head home that's all, I've got stuff and work and—"

"Maya, slow down, I'm sorry. I must have misread. At least let me call you a taxi."

"That's fine, Ethan, I'll get the tube."

"Maya—"

I left the house and shut the door behind me.

I thought about what Naomi and Tim would ask when I didn't

come back outside, about what Ethan would tell them. What would they think of me when they heard?

I walked up the street, taking pacy, shallow steps. When I got to the main road, I watched the cars go past like small, colorful bugs. It could have been ten minutes or thirty minutes that I stood there—I had no idea. As soon as I saw a black cab with an orange light, I hailed it. It pulled over just as I was starting to feel dizzy. I got in.

"Shepherd's Bush, please."

I knew this would cost me, but I wanted to get home as quickly as possible.

"No problem," said the taxi driver, indicating and pulling out into the busy road.

I sat in the cab for half an hour, replaying the kiss over and over in my head, thinking about what I'd said, what I'd done, and asking myself why. Why had I done that? I felt the buzz of the insect in my stomach, running through my blood and my brain.

I got home, paid the driver, and crept into the house. All the lights were out. My parents must be asleep, I thought. I removed my heeled boots, got a glass of water, and tiptoed up to my room in the attic. After opening the skylight, I lay on my bed, fully clothed, and eventually I checked my phone. There was a message from Ethan.

> Maya, I'm so sorry. Let me know if you
> want to chat tomorrow x

I also had one from my mum about an hour before.

> Are you coming home tonight? Just let me
> know you're safe xx

I replied to Mum.

Home safe xx

I left Ethan's message unanswered, then switched my phone off, put it on my bedside table, and turned off the lamp. As I lay there, my head spun. Lights seemed to flicker, tube lights fluttering, and the girl sat there, reading her book.

6

FINALLY

I WOKE UP the next morning with a faint hangover. Head throbbing, eyes aching, dry throat. I raised my head, looked down at my body, and groaned before plonking my head straight back down. I was still fully clothed in the outfit I'd worn last night, and I stank of alcohol and cigarettes.

An empty water glass lay sideways on the bed, and there was a stain like a faded watercolor painting on the sheet. I sat myself up and looked at the makeup remnants on my pillow. I felt so thirsty. There was a half-full bottle of water on the bedside table, so I grasped it and sucked on it like a lamb until it crunched into a shriveled plastic shell.

My head felt full of helium, squeaky and light, and there was a tender spot between my boobs where the insect was about to be reborn.

I had to keep moving, I told myself. Scrambling out of bed, I went to open the blind. I could only reach it by kneeling on top of my creaky wooden desk. Light burst into the room as I yanked it up.

I peeked out onto our overgrown, daisy-filled garden. The sunshine stung my eyes as I observed the occasional butterfly or wasp whiz past. Sounds of London droned in the background: the rumble of the tube, the hum of red buses, the beep of black cabs, the shouts and laughter of people on the street.

I got off the desk, removed all my clothes, put them straight in my washing basket, then wrapped my towel around my body. As I walked down the stairs to the bathroom, I could hear my mum in the kitchen. She was on the phone to someone, laughing.

Door to the bathroom shut, shower on, towel hung up, I briefly looked at my dark brown eyes and angular face in the mirror, then at my body. I cringed at the bumps and curves. I resented the contrast between how I looked now compared to my polished appearance the night before. My hair was becoming curly again, and my dissolving makeup revealed the plainness of my facial features. I clasped my eyes shut and turned away from the reflection, opening them again to get into the steaming-hot water. Scrubbing my face, washing my hair, cleaning under my nails, all of it was so soothing. My feet twiddled on the fluffy bath mat as I got out the shower and wrung out my damp, frizzy hair before heading back to my room. Shutting the door to the bedroom, I stood there.

My stomach continued to twitch. I just wanted to get back into bed and do nothing, but I had to keep it together.

After a moment, I reluctantly picked up my phone from my bedside table and turned it back on. I sighed, sat on the side of the bed, and waited for the messages to load.

Naomi:

> Are you ok? What happened with you and Ethan? Get home safe and hope you're alright. Message me in the morning xx

Ethan:

> Maya, I'm sorry again. Please let me know you're not mad at me. Did you get home ok?

I sat there staring at the screen, then typed back a message to Ethan.

> Don't be silly—I'm not mad. Sorry I freaked out. Shall we talk when I see you at Lucy's birthday tonight? X

I perched on the side of the bed a bit longer, letting myself drip dry and feeling the wetness of my hair against my back. As I sat there, I tried to understand myself, but I couldn't answer any of the questions spinning around my head. I asked myself why I had felt so overwhelmed and panicked after Ethan kissed me. This was Ethan. I'd thought about this for so long, but something inside me was holding back. I shouldn't be holding back; I wanted this. If something happened between us, surely everything would slot into place. Perhaps this could be the thing that was missing.

I'd never had a boyfriend before, not properly. I'd kissed a few boys at school, and during my first year at Edinburgh, I'd slept with one guy. He was in his final year, and I'd met him through poetry society. It only happened once, and I hadn't even enjoyed it really. It had been painful, and I'd mainly done it to get it over with.

This was different. It meant more to me than anything had done before. I'd known Ethan for so long, and things would start at a much more serious level if anything happened between us.

My phone buzzed.

Definitely, let's talk later. I'll see you there.

I stared at it, my head spinning, my stomach contorting. Two seconds later, my phone vibrated again.

Keen to see you x

• • •

LUCY, WHO HAD been at school with Tim and me, had her birthday party that night at a bar in Soho. The place was already crowded when I got there. I'd put myself together for the evening. Makeup, hair clipped up, satin dress on.

I arrived alone. Normally I hated doing that, but I felt too nervous to see Naomi or Tim before the party. I didn't want to give them an explanation about my swift disappearance from drinks the night before until I'd cleared things up with Ethan.

As I pushed my way through the sea of bodies, the music pumped out of the speakers. The bar had mirrors behind a wall of bottles, and they reflected the alien greens and sherbet yellows of the acidic liquids.

When I saw the group, Naomi bounded over to me and hugged me tightly, whispering something in my ear about last night that I couldn't hear properly. Tim also came over and put his arm around me. Then I saw Ethan between the gap of two other people. He was wearing a navy linen shirt and beige trousers, with his hair smoothed to one side. His skin seemed to glow as if it had been freshly shaven and moisturized. As we made eye contact, he tipped his head slightly, smiling at me. He looked at me as if to ask if we were OK. I gave him a slight nod, and then went to find Lucy. After queuing for ages to buy her a drink and going back to find the group, I felt a light touch on my arm and warm breath in my ear.

"Wanna head outside for a bit?"

I turned to see him and felt my body melt as he grasped my hand and led me to the smoking area. Naomi was out there. She had one arm crossed over her body and was inhaling deeply from her cigarette, making sleepy eye contact with some guy who was borrowing her lighter. Ethan steered me away from her, towards the opposite corner. He took out a filter from his trouser pocket and stuck it in the corner of his mouth.

"I'm sorry about last night. I'm embarrassed. I thought I was getting signals from you recently, but I must have misread. Are we good?"

He took the filter out of his mouth and placed it on the paper, then shuffled the tobacco around and rolled it into a cylinder with the tips of his fingers. I watched him lick one side to stick it together. His tongue was pink and moist, like the tip of a raspberry. Once he'd smoothed it out, he tucked the cigarette behind his ear and leant against the wall, his body close to mine.

I stared at him. As he'd been speaking, I'd been thinking about the texture of his hair and the softness of his skin, about what his teeth and tongue would taste like inside my mouth. His eyes glinted as his face burst into a grin. A grin I knew so well.

I rolled my eyes as he wriggled with glee, readjusting the cigarette behind his ear.

"You really are terrible, did you know that?"

"Awful." He smiled.

"I mean"—I flapped my hands helplessly—"you've been giving me signals, too."

"I have." He nodded.

"So, you, what?" I left a gap and moved my hands around in the air.

"I've liked you for a while, Maya," he laughed. "Haven't you noticed?"

I felt completely weightless. People had joked about Ethan and me getting together before, saying things like "You'd be such a perfect couple" or "Your children would be so cute," but I never actually thought it was real. I never thought this could happen.

We stood there for a minute, our figures fidgeting. We were leaning against the wall, our faces close.

"And what about our friendship?" I contemplated in a small voice.

"Yeah. No, you're right." He shook his head and stared down at my feet. "It would be really stupid."

He tilted his head back up to meet my gaze, and we waited, still and peaceful amid the movement around us. I breathed in.

This time when he kissed me, I kissed him back.

It felt right, being with Ethan. As we kissed, I told myself that was why I'd freaked out the night before; I'd known how significant this was going to be, and I didn't want to be a disappointment.

He drew me into him, and my stomach flittered, like a feather tickling at the base of my pelvis.

• • •

WHEN I WOKE up on Sunday morning, I was alone in Ethan's bed. I turned over into the empty space and saw a small note on his pillow.

I've gone to catch my flight to France. Message me when you wake up. Last night was amazing x

I read it over three times, smiling to myself.

It was a hot day in London. The rays of sun pulsed in through the window, dampened by thick blue curtains. I opened them and looked down onto the quiet suburban street. The town houses stood like men with mustaches, gloriously arrogant.

Ethan's room felt so fancy. Cream carpets, grand lamps, wooden floor covered by a Persian rug. On top of it all was the mess of the strewn clothes and random bottles of moisturizer and deodorant.

Intimidated by the creaking quietness of the house, I put on my clothes, checked that all the doors and windows were locked, then left swiftly to catch the tube home. As I sat there in the mugginess, I removed my jumper and felt my stomach croak. I shut my eyes. There it was again. The insect chewed at my flesh inside and spat it out into a heavy lump.

Walking home from Shepherd's Bush station, I desperately tried to think about other things. I rubbed my eyes hard with the bottoms of my palms. I could smell barbecued chicken, the waxiness of lemons, the freshness of ripe strawberries and peaches stacked up at the markets. People bustled past the kebab shops and the launderettes, past the bulging red and green bin bags stacked up on the pavement. Some hobbled with walking sticks, or strode with dogs on leads, or pushed babies in buggies. Some stood on street corners, smoking with their sliders off so their bare feet were burnt by the dark concrete of the pavement.

I turned onto my road, rummaging for my keys in my bag, finding them just as I reached the gate.

My parents' house seemed dim and cool, tinkling with the sound of jangling bamboo wind chimes. I ambled through to the kitchen, where knickknacks were stacked up precariously. My mum was an artist, a painter, and I would say her style was eclectic. Copper saucepans hung up above the gas cooker on metal hooks, plates with cerulean blue drawings on them, mismatched teacups, sunflowers bought from the market a week and a half ago, drooping over the side of the vase with lifelessness. Their petals dropped off onto the table like flaky autumn leaves.

The back doors opened out onto the lawn. I put my bag down. Looking out to the garden, I saw my parents sitting on chairs beneath an ivory parasol. There was a large copper cafetière full of coffee and the Sunday papers strewn across the table in front of them. I walked into the sunshine, putting my hands above my eyes.

"Maya!" my mum exclaimed, removing her spectacles. "I was just about to send out a search party."

"I let you know I was safe, though, it, I was—"

"We won't ask where you were," my dad interrupted, nodding towards my mum, then giving me an almost imperceptible wink. He hadn't removed his glasses and immediately went back to reading the paper.

"Thanks," I sighed.

My mum waggled her bottom in her seat. I could tell she was desperate to know where I'd been and who I'd been with.

"Do you want some coffee? Your dad just made a pot and I think the culture section of the paper is up for grabs."

"Thanks. I might just shower and then come down. Are we still going to that exhibition later?"

My mum shrugged warmly. "I'm up for it if you are. Around two?"

Dad continued to read the paper, a contented look on his face. "Sounds good. I'm working at the pub from seven 'til close."

Mum smiled at me and nodded.

Up in my room, I opened the skylight wide and watched the pollen float through the thick air. I tried to breathe away this feeling in my body. I tried to let the insect escape.

7

PAINTING

AUGUST 2013

ONE MORNING AT the end of August, I stood in the garden with Mum. The day was bright and warm. She wore a floaty dress, and I had on loose shorts and a strappy top, my hair encased in a silk bandanna.

Over the summer, we'd made a routine of putting two easels up in the garden if the weather was nice. We'd stand there for hours, side by side, painting.

That morning, she was working on an oil on canvas based on a photograph of a sunrise over a field. I was attempting a watercolor of a bowl of fruit which rested on a small table in front of me. My painting wasn't very good, but Mum was teaching me various techniques and I was slowly improving.

Usually I preferred to draw. Drawing and reading poetry were my two favorite pastimes in the whole world. I'd almost applied to go to art college after school, but I'd decided that maybe I could do that after my English degree if it was still something I felt passionate about. Dad had really encouraged me to do the English degree first. He wanted me to have a backup, he'd said.

I had done quite a bit of drawing as well as painting that summer, but the drawing wasn't something I'd done in front of Mum, or in front of anyone. I usually drew women. Revealing the parts of their bodies that were hairy, smooth, dry, irritated, bloody, curved, flat, lumpy, tight, delicate, heavy, pierced. One day I decided to draw the girl I'd seen on the tube that time.

I used charcoal. I drew the way her leg muscles showed through her linen trousers. The way her breasts hung under her floating turquoise top. The way her hair sprang out of her headband. The light revealing her and casting her shadow.

When the drawings were done, I hid them under my bed beneath some boxes of old schoolbooks.

The summer had been a complete whirlwind, but in a good way. I'd had a part-time job at the pub down the road, which I enjoyed. I liked chatting to people, and the atmosphere was always lively. I usually worked three long shifts a week, sometimes four.

When I wasn't doing that, I'd been spending time with Ethan since he got back from his short trip to France. If it was sunny, we'd sit in his garden or in mine. I'd read a book and he'd do some of his vacation work for Cambridge. Sometimes he'd lean over to look at what I was reading and tell me he didn't understand poetry. I'd laugh at him and push his face away, telling him to stop being ridiculous.

As expected, our relationship had become serious from the get-go. My parents found out quite quickly. My mum had actually guessed it, and she'd looked so happy when I confirmed her suspicions.

"Oh, I do like Ethan." She'd clapped after I told her. "He's such a lovely, clever boy. I just knew you'd end up together."

Of course Naomi and Tim knew, too, as did all our other mutual friends. Just as they did before we even started dating, people kept telling us how we made such a great couple. I felt as if I'd gone up in people's esteem, and that made me feel almost proud.

Ethan and I had stayed together most nights over the summer. We'd go to the pub or have dinner with Tim and Naomi and then stumble back afterwards to one of our houses. Sometimes we'd just cook dinner at his, putting on loud music and dancing around, drinking lots of wine. His parents were away on holiday quite a bit, so we often had the house to ourselves. If his house was empty, we'd have loud, drunken sex when we got home, tasting the alcohol and tobacco on each other's tongues.

"How is everything with Ethan going then?" my mum snooped as we stood in the garden that morning, painting.

She peered over her thin rectangular glasses at her canvas, doing little teal streaks at the base of the sky. Her hair was in a long, dark plait down her back.

I smirked. "Mum."

"What?" she hummed. "Am I not allowed to ask that?"

I didn't respond. I just kept trying to mix the shade of red I wanted. I couldn't get it right and was beginning to feel frustrated.

"We like him, your dad and me, that's all. He's a sweet boy."

My parents had known Ethan since we were young, and they'd always hinted to me how much of a nice person he was. He'd always been the sort of guy to walk me home or take care of me when things happened. When we'd been sixteen, I'd gotten so drunk at a party that he'd had to call the ambulance, and then call my parents. He'd even gone with me to the hospital where I'd got my stomach pumped. My parents had talked about him for weeks after that incident, saying how grateful I should be to have a friend like Ethan. Truth be told, I was grateful.

After we started going out, Ethan had come over for family dinner. I could tell that both Mum and Dad loved having him there. He dressed smart and brought a bottle of wine with him. Generally, his manners were impeccable. He was helpful, attentive, and even made my dad laugh at the table, an achievement of the highest magnitude. I loved my dad; he was a gentle person who worked hard and treasured his family dearly, but he was not the sort of man to hand out a chuckle easily.

"I'm glad you like him," I said, finally managing to mix the correct shade of red.

"Yes, well, of course it makes me pleased to see you happy."

She stepped back to look at her painting, then at mine. I could feel her assessing it, analyzing each bit.

"Very nice, sweetheart, you're getting good at this painting thing. Might put me out of business soon." She laughed. "Have you thought about bringing out the contrast a bit more around here, though?" she said, pointing towards the color of the apples.

I stepped back to review my own piece. The fruit, the bowl, the table. The futile imitation of reality I'd created. I sighed.

"You're right, I'll work on those."

• • •

EVEN THOUGH IT was only the end of August, it was almost time for me to head back to Edinburgh. Annoyingly, we started at the beginning of September, whereas Ethan didn't start back at Cambridge until early October. Although I was itching to return and see everyone, I wanted longer with him. It felt as if our time was being cut short.

A couple of days before I had to leave London, Ethan told me to meet him at Hampstead Heath. He had a surprise for me. Along with his message, he'd sent me an annotated map with an arrow pointing to an exact location. I hated surprises, but I humored him and said I couldn't wait.

I straightened my hair and clipped it half back, putting on lipstick and perching my sunglasses on my head. When I got there, people were dotted about on the lush grass, lying underneath the elderly, dripping trees, or sitting by the edge of the murky water. The sky was setting into a mellow apricot shade, caressed by icy blues and candy floss pinks. Dragonflies glowed, dancing like fire, and the pigeons cooed. Children's kites danced in the sky like birds.

I spotted Ethan lounging next to the big pond. He gave me a small wave. His skin was olive-colored from the summer, his chin slightly stubbled. Next to him there was a picnic basket.

As I sat down, he kissed me.

"This looks amazing. Is it just for us?" I smiled, removing my coat and peering into the basket.

"I just thought before you left it'd be nice to have a romantic evening." His face spread into a grin as he retrieved a bottle of wine.

I watched him as he uncorked the bottle and poured the liquid into two plastic glasses. He handed me one, and we clinked them together before taking a sip. He looked deep into my eyes as he swallowed, and I felt my cue to do something. I put my palm on his cheek.

"Thank you."

"It's nothing," he said quietly, leaning his face into my hand. "Maya, this summer has been—"

"It's been incredible."

"I'm so glad Naomi is shit at setting people up." He smiled.

I laughed awkwardly and tipped my head sideways, watching Ethan stare at me. He looked at parts of my face. My nose, my lips, my eyelashes, my earlobes.

"I really like you," Ethan whispered.

My stomach jolted as I responded. "Well, that's a relief; I quite like you, too."

He nodded seriously, as if we were having different conversations.

"We'll make it work, this term. I'll come up and visit you and you can come down to me, maybe we can do one weekend trip a month, that way we see each other every fortnight but it's not too much traveling for either of us," he said, taking my hand in his. It was clammy.

I dropped my smile and put down my wineglass, clasping onto him. I thought about how grown up this was, being sat here with Ethan, vowing to make things work.

"That sounds perfect." I nodded.

He leant his face forward and kissed my hand.

We sat there for hours as the sun set into a deep blue haze. We ate crusty bread and punnets of ripe strawberries. We drank a lot of wine. We talked about how much we liked each other, reviewing and dissecting all our missed opportunities over the years. Our laughter filled the air around us, carrying into the tangled branches of trees. As it got even darker, I lay down and rested my head on his stomach. We looked up to the sky. It was clear in some places, like see-through lingerie. I could make out a few glimmering stars.

"I can see a triangle of stars there," I said, pointing upwards. "What does that mean?"

"I'll look it up. I've got an app which tells you."

I kept my eyes on it as he got out his phone and held it up.

"The light pollution is so bad," I muttered to myself.

Ethan cleared his throat and read out the description.

"Apparently that's the Leo star sign. It's made up by the bright stars of Regulus and Denebola, the nearby star Wolf 359, and several famous deep-sky objects. Apparently, it's the twelfth largest constellation, occupying an area of nine hundred forty-seven square degrees."

Looking at how small these stars seemed to me, I thought about their actual size and how, really, I was so small in relation to these huge balls of hydrogen and helium. I thought about the fact that in our Milky Way galaxy alone, there are estimated to be around three hundred billion stars, and nebulae are, even now, birthing even more stars.

I felt Ethan's stomach inflating, then deflating with his breath. Peering upwards, I thought about how many stars there were up there that I couldn't see. Stars that were hidden by distance, or time.

Part Three

Encounters

8

NEW LIFE

Aisling

County Clare, Ireland → Edinburgh, Scotland

SEPTEMBER 2013

I FLEW FROM Shannon to Edinburgh on a drizzly September morning.

Ma dropped me off outside the airport and stood there while I got my two suitcases out of the car boot. One of them was full of bedding and towels; the other contained my clothes and shoes. I also had a backpack stuffed with my laptop and a ton of books. Pa hadn't come with us. He hadn't even said goodbye to me; he'd left for work before I woke up that morning.

After I'd got my luggage from the car, my mother and I stood quietly on the trout-gray pavement, looking at each other. Things happened around us. Planes hoovered up the sky above; cars came and went, beeping and braking and pulling out. People hugged and kissed or ran hurriedly or walked slowly around us.

I realized how similar we appeared to be. It was like a mirror sometimes, looking at her, but her face also became less familiar to me in that moment. It was almost as if I were preempting the distance between us. I looked at her faintly wrinkled skin, the wispy hair sticking out from behind her ears, the sunspots running along her cheekbones. I thought about all the words that had passed over her lips. All the horrid things she'd ever said and done to me. I thought about everything her eyes had expressed to me when I'd hoped for love and acceptance.

"Safe travels then," she mumbled.

"Thanks. I'll see you at Christmas."

Ma nodded and gave me a faint smile. After a moment, she abruptly pulled me into her body, her limbs all at right angles like a cardboard box. Then, suddenly, she grabbed my upper arms and pushed me away from her, got into the car, and drove off without saying another word.

On the plane, I felt nervous. The air-conditioning spouted out of the vents. The air hostesses' grins were surrounded with cherry-red lipstick as they came to check the seat belts. I couldn't have felt further away from their projection of geniality. All I had wanted when I was home was to be alone, to be far away from my parents, and now I felt sick at the thought of being separated from them.

I recalled, in that moment, something Orla had once said to me, when I'd confided in her about my mixed feelings on leaving home. I felt so dependent on my parents, even after everything.

"Have you heard of Stockholm syndrome?" she'd said to me, itching her upper arm.

I'd shaken my head.

She'd got out her phone and read the description to me: "'Stockholm syndrome is a psychological response which occurs when hostages or victims bond with their captors or abusers. This can develop over years of captivity or abuse.'" Her words had trodden slowly and carefully, her green eyes peering up at me like a still pond.

I reminisced about this moment as I sat on the plane. I hadn't spoken to Orla since Sean had caught us; I knew she wouldn't want to talk to me again. She hadn't even reached out to me to say happy birthday. We'd got our Leaving Cert results a couple days after my birthday in August, and I'd seen on social media that she had gotten into Cork. I wanted to message her and tell her I'd made my offer for Edinburgh, too, but I'd stopped myself. There was no point; we were both going to start anew, so why drag up the past?

I took a long breath. The buttery smell of the warm croissant I'd bought in the terminal seeped through the oil-stained brown packet. I was scared about going to Edinburgh, sure, but it was good. I needed to do this. I had to see what life was like beyond my current existence.

As we took off, I slowly ate the pastry, which flaked onto my lap. I breathed deeply and turned to look out the window. The fields joined together like a rolling quilt. Shades of olive, emerald, mustard, lined by speckles of trees and bushes. Morning sun cracked through the clouds and illuminated the land with streaks of light. I spent most of the journey staring out at the crystal-blue sky which had been hidden before.

There were no direct flights, so I had to change in London, but eventually I got to Edinburgh Airport. I loaded my suitcases onto the bus and caught it into town. I'd never been to Edinburgh before, only seen it in photos before deciding to apply. At the time, I'd liked the idea of being divided from my parents by a body of water.

It was beautiful, the city: the uniform rows of smoke-stained town houses, the purple hills and crags towering in the distance, the grand castle overshadowing the streets, the churches and monuments, the winding cobbled alleyways, and the promise of the sea beyond and beneath it all.

Getting off the bus, I checked the map on my phone to see where my building was exactly. My hands were shaking. It was just beyond the meadows: a broad expanse of green with tree-lined paths which lay below the looming hills leading up to Arthur's Seat.

I dragged my suitcases along as the light faded. The air seemed fresh from the rain. I could smell the dew-stained grass, the damp air, and beer hops.

Once I'd registered and got my keys, I headed up to my flat. I was meant to be sharing with two others, but the place seemed empty and eerie. I walked into my room, which faced the back gardens, and I stood there with my two suitcases next to me, my backpack still on. The flat had cream walls, blue carpets, and tall windows which looked out onto skeletal trees. I let the door close behind me and lingered there in the darkness, listening to the wind and the slow drips of rain outside. I was still somehow shaking with nerves. Sean, Mary, Jack, Ma, Pa, they all felt so far away from me now.

I opened the windows, unzipped both suitcases, then made the bed with the sheets I had brought. After, I unpacked all my clothes, books, and toiletries. Observing my room, with the few things I had filling only a small part of the space, I felt so alone.

My stomach groaned and I realized I hadn't eaten since the croissant on the plane that morning. I'd seen a shop right across the street, so I grabbed my purse, left the apartment, trotted down the stairs, and headed across the cobbled road. It was dark out now.

In the store, I wandered around the perfectly stacked shelves of granolas and mueslis, strawberry jam jars, olives floating in brine, past the sleek fridges full of alcohol. I grabbed a basket and plucked a few things off the shelves.

When I got to the checkout, there was a girl standing in front of me in the queue. She had long legs and curly hay-colored hair. In her arms, she cradled a loaf of bread, milk, butter, and eggs.

"Ayte, love," the man behind the counter greeted her.

She said hello and dumped her items on the counter. As he scanned, she drummed her card against her manicured fingernails. When I went up after her, I didn't utter a word to the man behind the checkout, but he smiled at me as I paid and picked up the groceries.

Back in the flat, the kitchen light was on. I walked up to the door and saw through the glass that it was the girl from the shop. She was at the kitchen counter, cracking eggs into a clear jug, drawing apart the shells like curtains. I took a deep breath and entered.

"Hi there," I mumbled.

She jumped slightly and turned, touching her chest.

"Sorry." She laughed at her own reaction. "I thought I was the only one here."

I placed my groceries on the counter. "Me too," I responded, grinning. Her eyes were small and green. Her lips thin but smiling.

"I'm Aisling."

"Karen."

Awkwardly, we shook hands and then chuckled at ourselves for doing it. She turned back to the counter and started to whisk her jug of eggs with a fork. I got out my ready meal and put it in the microwave, which beeped as I pushed the buttons. It started to purr as it rotated. I'd realized while standing in the shop that I didn't have the faintest idea how to cook, but I'd somehow been assigned uncatered accommodation, so I'd soon have to learn.

"What're you studying?" she asked me as she put a pan on the stove.

I cleared my throat to cover the shakes. "English and Scottish literature."

"Nice. What's your thing?" she asked.

"My thing?" I responded, raising my eyebrow.

She giggled in a self-deprecating manner. "I just thought, like, English students always seem to have their favorite writer or genre or whatever. But I don't know." She shrugged and flailed her hands about. "Maybe that's silly."

"Not silly," I reassured her, closing my eyes. "I guess I like poems."

"Poems," she repeated.

I opened my eyes again. I was trying to get rid of the dizzy feeling in my head.

"Yeah, poems. Reading them, I mean. I'm no good at writing."

I had been attempting to do more writing over the summer but had decided I was lousy at it. The only one I'd finished was about Orla, and it wasn't any good.

Karen poured her eggs into the hot pan. I could see her tipping her head to the side as if assessing something.

"Have you heard of the poetry society?" she asked me.

The microwave pinged.

"What, like the film?"

"No, that's *Dead Poets Society*," she laughed. "My friend Harry, we're both from Glasgow, he runs this society here at Edinburgh. It's part of the university, it's for people who enjoy poems, I think. They have, like, weekly discussions and stuff."

I went to get my food from the microwave. "Oh, nice."

"I'll message him and send you the details. Maybe you could join?" She shrugged.

I nodded to her as I sat down with my meal, wondering if I'd ever be brave enough to go to something like that. Surely everyone would know what they were talking about.

"Thanks," I whispered. "I'd like that."

That evening, Karen and I sat in the kitchen for hours, just chatting. It felt nice to speak to someone my age. Karen told me more about where she lived in Glasgow, and that she was in her second year studying geography. She'd been planning to move in with her boyfriend that year, but he'd broken up with her just before the summer. I told her that was rough, but hopefully I wouldn't be a terrible replacement as a housemate, which at least made her smile. She asked me more about where I was from, about my family. I gave shallow answers, mentioning my parents and siblings only in passing, avoiding anything too detailed. Slowly, as we spoke, my nerves started to dissipate. I even began to feel something like excitement, as if there were a new life ahead of me.

• • •

THE FOLLOWING WEEK, on the Thursday evening, I showed up at the poetry society meeting. In my bag I'd packed a notebook full of my attempted drafts of poems, a book of actual poems by Eavan Boland, a bottle of water, a pencil, and a sandwich.

Karen and I had been joined by our other flatmate earlier that week, Toby. All I had learned about him was that he played some kind of sport quite enthusiastically. I wasn't sure which sport, but he usually showed up after training at around six thirty p.m. and ate the least appetizing mixture of stinking meat and dairy, so I tried to avoid the kitchen at these times. The poetry society meeting was scheduled to go on until six p.m., so I had packed the sandwich in my bag and was planning to eat it on a bench outside afterwards in order to avoid the flat.

I'd decided to come to the meeting on a whim but regretted it as soon as I got there. There were about ten people in the room already, and more arriving gradually. I genuinely had no idea so many people liked poems. All of them nattered away to one another, exchanging drawn-out hugs and speaking in exaggerated tones about how pleased they were to be back in Edinburgh. I found a seat near the back and sat quietly. I took out my notebook and pencil, along with my Eavan collection. I felt so nervous, almost like I might vomit. I thought I should just get up and leave. No one had seen me anyway. Slowly I put my hands on the books, placing them on my lap and turning in my chair, ready to make a swift exit.

"Surely you're not leaving?"

I looked up and saw a girl in front of me.

"I promise we're not that bad," she said, smiling warmly, as if greeting an old friend. Her hair was scraped back into a poker-straight ponytail. She wore large hoop earrings, a floral headband, a sky blue jumper, and flared trousers. Her eyes were an earthy brown. I felt the air being squeezed out of my body, my chest and neck flushing red.

"I'm sure you're not," I said.

"Good, so you're staying."

She spoke to me as if she'd known me for years. I couldn't see the people around us anymore. Only her.

"I love Eavan Boland's poetry, by the way," she said gently.

I peered down into my lap to see the book there, then caught her eye again. I couldn't stop myself from smiling in disbelief.

"Do you now?"

"I do, yeah. Is this seat taken?"

The girl pointed to the empty chair next to me. It could not have been emptier. I shook my head. I could feel the sweat on my back. As she sat down next to me, she unpacked her bag. A silver water bottle, a notebook, a pen.

"Maya."

"Huh?" I breathed.

The girl turned to me. "I'm Maya."

She smiled, the corners of her mouth sending shallow wrinkles into her smooth cheeks. Turning back to face the desk, she opened her turquoise notebook. I stared at her. Like a polished stone, her skin looked almost velvet to the touch. Thick, ferny eyebrows curved like a soft accent on a word. Her lips were the color of coral.

"I'm Aisling."

"Aisling—like the genre of poetry?" She paused. "Isn't an aisling a dream or vision poem or something where Ireland appears as a woman? A sky woman or heavenly woman or . . ."

Maya squinted and turned to face me. Catching her eyes made it difficult for me to swallow. I looked forward, away from her, to regain my nerve. All I could feel was my heart moving in my chest.

"A lamenting woman hoping for better things for Ireland probably sums me up pretty well, to be fair."

She laughed. It sounded like a waterfall hitting a lake. "Doesn't Heaney or Muldoon or someone have a poem called 'Aisling'?"

I laughed then. "They do. I think that probably sums me up well, too, to be fair."

"Oh yeah?"

"Deeply ironic," I mumbled, turning back to her with a smirk.

Her smile glistened as it spread across her face. We held eye contact for a moment, and then looked away from each other.

I glanced over to her hands on the desk in front of her. She was scribbling a drawing of something in her notebook. It looked like an angel.

"OK, everyone," shouted a lad at the front of the room. He wore a polo shirt about three sizes too small for him.

"Time to start. Welcome to all the new people. Let's just go around the room and say our name and what we're studying. I'll start. I'm Harry, I'm the president, of poetry soc, obviously, and I'm studying history."

We went around the tables. I tried to remember everyone's names. There was Isla, Tommy, Ruby, Rohan, Isabel.

I had always hated speaking in front of people. Mam had always forced me to read at Mass for all those years, and I never got the hang of it. As the person before me finished introducing themselves, the room focused on me.

"I'm Aisling." I paused to clear my throat. "I'm from County Clare in Ireland, uh, and I'm a first-year, doing English and Scottish literature."

The room nodded. I breathed out. My whole body sank into relief. I turned to Maya, and it sank even further. Her eyes latched onto mine; then, after a moment, she turned out to face the room and leant back in her chair.

"I know lots of you already. I'm Maya. I'm doing English as well, and I'm in second year." She started to fiddle with her pen. "Oh, and I'm the social secretary for society this year, so I'll be bribing you all to join me at various drinking establishments."

The whole place rustled with laughter.

There was something about her, something which drew me to her but which also felt completely untouchable.

9

DRAWING

Maya

I TURNED THE keys to the door of my new flat.

It was one of those perfect Edinburgh days in early September where you just about don't need a coat. The sun poked out from behind the cotton clouds, and the breeze off the hills was soft.

My friend Gabe and I were renting together for this academic year, but he wasn't set to arrive until the following day. Gabe studied law, and he had an unruly mop of curly brown hair and big tortoiseshell glasses. He and I had met at the start of first year through a mutual friend from London, and I had immediately taken to him. He had this calm, caring aura, but at the same time he was extremely fun.

As the door to the flat swung open, I smiled. Everything seemed to be falling into place, I told myself. I was back in my beloved Edinburgh, I'd had the best summer with all my friends in London, Ethan was my boyfriend, and I was sharing a beautiful flat with one of my best mates. I walked inside, dragging a couple of suitcases behind me. The place was exactly how I remembered it when Gabe and I had seen it the previous year. Georgian windows, high ceilings, beige carpets, whitewashed walls, pine furniture. The corridor led off to the kitchen, the bathroom on the left and the sitting room on the right, with the two bedrooms down at the end.

I walked straight towards my bedroom, the one which faced the street. After a moment of debating whether I should re-arrange the furniture, I went over to the window and opened it to let in the breeze. It was clear and sweet, a different kind of air than in London.

I heard Ethan entering through the front door, shouting my name.

"I'm in here!" I yelled.

Given that Ethan didn't start back at Cambridge until October, he'd driven me all the way up to Edinburgh using one of his parents' cars.

Doing the journey all in one go would have been too long, so we'd stopped off in York the night before and got a hotel room. In the evening, we'd walked around the city in the rain, huddling our bodies close together under a large black umbrella. The slopping streets and the medieval stone of York Minster all seemed quieter in the lamp-lit dusk. Afterwards, we went to an Italian restaurant for dinner, ordering pasta and a bottle of wine. Ethan only had one glass because he didn't want to be hungover for the drive the next day. I drank the rest of it. I felt quite drunk when we had sex later that evening. The whole thing had been hazy, and I hadn't come, although I rarely did. Ethan had moved my body around, telling me where he wanted me and what I should do. While we lay in bed afterwards, we chatted about our plans for this term, studying each other's fingers as we twisted our hands together.

Ethan emerged into the bedroom of my new flat laden with bags. He deposited them on the floor.

"That's fine, Maya, I'll get all the bags."

"Sorry," I said, moving towards him.

"How have you got so much crap?" he laughed.

He straightened up and rolled his neck and shoulders around. After he'd assessed the room, he took off his cap and threw it on the bed, walking closer to me still.

"The place is nice. Are you happy?"

I nodded, casting my eyes around.

"When did you say Gabe was getting here?"

His hands caressed my face, and gently he tugged me into a kiss. He tasted of cigarettes and coffee.

"Tomorrow," I said quietly, wiping my lips.

"Very interesting. So, we have the place to ourselves?"

I laughed and pulled away, lightly touching his stubbled cheek, then turning towards my bags and starting to unzip the cases.

"Would you be able to go out and grab some food so I can start unpacking?" I muttered.

There was silence behind me for a moment, and I looked around to see Ethan's puzzled face.

"Sure," he croaked, checking his pockets for his wallet and cigarettes. "I'll ring up when I'm back. Any preferences?"

"Just some milk, butter, and bread, then whatever you fancy for dinner."

"Sure."

"Thank you."

I heard him leave the room and the front door slam shut behind him. I stayed there, kneeling on the floor, looking at all my belongings in front of me. Everything seemed quiet. All I could hear was the faint sound of traffic in the background, the distant tinkle of the streets below. Everything was perfect, I told myself. But the truth was, I could still feel it. I could feel the insect, like a wasp inside of me, ready to sting.

• • •

ETHAN STAYED FOR the next couple of days before term started. The morning he left, he kissed me and told me he loved me. "Same," I replied. I felt sad that he'd left, but it felt good to get on, sort out my stuff in the flat, and dive headfirst into term.

That week there was the first poetry society meeting of the year. On the day of the meeting, I had lectures in the morning. At lunchtime, I went to my favorite bookshop in town, just down past Grassmarket. It smelled of leather and old paper. Worn and lopsided rugs covered the floor, and the walls and overhead shelves were crammed breathless with books. Colors merged into one another like paints on a palette. I wandered around aimlessly, reading blurbs and first pages.

At the library that afternoon, I lost track of time. When I looked at my watch, it was quarter to five. I only had fifteen minutes until the meeting started. I packed up my books and pens, went to the loo, filled up my silver water bottle, and walked briskly down the road.

When I arrived, I saw familiar faces. Isla, Isabel, Rohan. There were waves and grins all around.

But before them all, I saw a girl sitting at the back of the room.

Her dark, silky hair caught my eye. It was half tied back, the bottom layer reaching down to the nip of her waist. She had two books in her lap and was spinning around in her chair as if she might get up and leave. The sight reminded me of myself last year. I'd been so nervous at my first meeting, and I wondered if she was, too.

When our gaze met, my stomach whirled, but not in a way it had ever done before. Her eyes were such a deep shade of blue, I felt as if I'd been immersed in water. They were encased in dark, thick eyelashes, and her lips were the color of merlot. Looking at her was like studying a drawing that I'd stored under my bed. It felt wicked and addictive.

Aisling, her name was Aisling.

We stayed for a bit after the meeting, chatting to each other. She asked me what kind of poetry I liked. I told her my favorites were June Jordan, Audre Lorde, and Jackie Kay. I asked her about where she grew up, about what poetry she liked, too. She was from County Clare, and then she pointed to her book by Eavan Boland and showed me her favorites. "In Her Own Image" and "The Other Woman." She read me the first two stanzas of one of them. Listening to her voice, I wanted to shut my eyes and let it envelop me.

10

COFFEE

Maya

"ALRIGHT THEN, WHAT do you want? I'm buying," Gabe grumbled through a cheeky smile. He cleaned his glasses on his checkered cotton shirt and replaced them on his face.

It was Sunday morning. Gabe and I were out for breakfast at a café near our flat, right by the meadows. It was our first outing together since we'd returned after the summer, our official celebration of moving in together. We were sat at a table right next to the steamy window. I'd unwrapped myself from my long coat and scarf; the temperature in Edinburgh had dropped even over the last week or so. The puffing and blasting sounds of coffee machines spraying water and gnashing coffee beans surrounded us.

"You know you're too good to me."

"I know," he sighed.

"Everyone probably thinks we're on a date."

Gabe rolled his eyes. "Do you want me to alert them to the fact you have a much more gorgeous boyfriend?"

I blushed, sighing into my lap.

"OK, I'm getting you a cappuccino. I'm bored of asking," he said, walking away to the glossy counter as he spoke.

"I'll get next time!" I called after him.

Gabe swatted a hand in my direction.

I took out my phone. I had one message from Ethan since we'd left the flat.

> Morning! How are you today? Missing you in London x

I put it in my pocket and gazed out the window.

On the other side of the road, about to cross towards the coffee shop, was Aisling. She wore a long, thick black coat and boots. Hair whipped across her face in the wind, but her elegant fingers strung it out of her eyes and tucked it behind her ears.

I had thought about Aisling every day since the meeting. I had ruminated on all those things that you never admit to anyone that you're thinking. I'd pondered what her family might be like, about whether she had a boyfriend. During the day, I thought about what she might be doing and where she might be sitting. I thought about her getting to know the other first-years, about whether she had been on nights out, about whether she had kissed or slept with anyone that week.

I'd added her on Facebook straight after we'd met, quickly regretting the keenness of it. Thankfully she'd accepted just the next day, which was a relief. I kept considering various excuses to message her, but I couldn't come up with anything good enough. I had looked through all her photos and even zoomed in on some of them, becoming familiar with the shape of her features. I found her face deeply comforting. The flick of her nose, the curve of her lips, the wideness of her eyes.

I'd been thinking about her so much that I kept expecting to bump into her, but for some reason I couldn't believe it was about to happen.

I felt panic rise in my chest.

I was hungover and was not presentable for anyone other than Gabe. I'd been out late last night and was sure I still smelled of alcohol and cigarettes. Even though I was wearing a bit of mascara, I hadn't had time to properly shower, straighten my hair, or put on perfume. I didn't want her to see me in these baggy jeans and slouchy jumper. I hastily tied my hair up into a bun before she saw me, rubbed to remove any makeup from under my eyes, and ran my tongue over my teeth.

As she walked towards the café, I couldn't help but watch her.

She opened the glass door and the bell jangled.

The fierce blueness of her eyes seemed to turn the air to ice, her blinks cutting sharply through the cold. Her head twisted and clicked, observing around the room like a small bird. Then she clocked me.

"Maya."

She was fresh from the outdoors, the skin on her cheeks slightly pink. There was an air of calmness about her. I was aware of my complexion, my body, my hair in contrast to hers. I could feel myself getting stickier with sweat, my cheeks getting warmer.

I asked her how she was.

"Alright, cheers. I live just around the corner so thought I'd see what the fuss was all about. My flatmate's always going on about the coffee here."

I nodded. Aisling looked down at the empty seat, Gabe's coat strewn over the back of it. She clapped eyes on it and then explored the room for my companion.

"Are you on a date? Agh, sorry, I'm interrupting," she said.

"Oh, no! No, no!" I babbled. "I'm here with my flatmate." I pointed towards Gabe, who was paying at the counter. "We live together, just over between Newington and Southside, so we're not far from here either."

Ethan ran through my head. I could mention him now, I thought, but then I decided against it. It would sound arrogant or off topic to bring up that I had a boyfriend just then.

"Well, sorry anyway," she murmured. "I don't want to interrupt or anything. I'll see you?"

"Wednesday!" I half shouted at her as she melted away towards the counter.

"Wednesday?" she repeated.

"We've got a social on Wednesday. For poetry society, if you want to join?"

"Yeah, grand, just let me know."

I could see her gaze flicking across my face. I sat there on my chair as she stood between two tables. For a brief moment, I felt that she saw me. Not just the outside of me, no, but something else. Then it broke, and she walked away with a turn of her lips.

As she joined the queue, Gabe came back to the table and put down the tray, offloading my cinnamon bun and cappuccino, followed by his black coffee and pain au chocolat.

Sitting opposite me, he said, "Gosh, you look a bit sweaty. Are you really that hungover?"

I touched my cheeks and my forehead. I could feel the dampness in my clammy hands. He narrowed his eyes slightly, then pressed his lips together.

I changed the subject quickly, asking Gabe to tell me all about his summer.

A few minutes later, I smiled and waved at Aisling as she left the café with her takeaway coffee.

"Who was that?" said Gabe. "She's gorgeous."

"Aisling," I muttered as she walked away, towards the meadows.

11

THE BAR ON LOTHIAN ROAD

Aisling

THE FOLLOWING WEDNESDAY, I walked into a bar on Lothian Road. Fruity perfumes and the scent of spilled beer wafted out the door as I clambered inside, past the crowds of people.

There was a buzz in the air. The clatter of glasses, of hands and pints smacking wooden tables, of people exclaiming, laughing, whispering. The smell of the sour alcohol hit me hard, reminding me of home. Apart from that, the whole scene was quite unfamiliar to me. Back in Ireland, I usually avoided bars or parties whenever I could.

This was the place Maya had told me to come to. On Sunday morning, I'd seen her at a café near my flat, and she'd messaged me shortly after with all the details. Earlier in the morning, on Sunday, I'd woken up with three missed calls from Orla. I still hadn't changed the name on my phone from "Joseph McMahon," but I found it liberating knowing that I could if I wanted to. Bit by bit, I was having the realization that I wasn't being watched anymore; I was free to do as I pleased. It was exciting.

In addition to the calls, there had been two texts from Orla that morning. Both were sent in the early hours whilst I'd been asleep. The first just said:

I miss you Ash.

The second apologized for the calls, saying that she was drunk and sorry.

I deleted both messages. She was probably in Cork, out with all her new mates. It might even have been a prank from one of them—how could I know? Anyway, I was trying to start afresh here, away from her and away from my parents.

After lying in bed for a while that morning, I'd got up, showered, and set off on a walk to clear my head. I'd strolled past the café and seen Maya through the window. All rumination on Orla drained from my mind, and in a split second, I decided to cross the road and go in. I had such a strong urge to see her. I had been feeling that way for days, ever since I'd met her. Seeing her had made my whole body feel a rush of adrenaline, and I left the encounter on a complete high, knowing I was going to see her again in only a few days' time.

The next day, I'd popped into the shop across the road on my way home from lectures. I walked through the aisle with the toiletries. It smelled like soap and washing powder. I went over to look at the makeup. The last and only time I'd worn makeup was when Orla and I went to the cinema the previous year. I shuddered with embarrassment when I remembered that I'd done just the one eye. But nonetheless, I picked up the eyeliner and mascara and walked over to the checkout, also grabbing a lavender candle which was on offer. Because, why not? This was exactly the type of frivolous thing Ma would never have let me buy, and so it gave me extra pleasure purchasing it.

"Not the usual then?" said Dave. Dave was the checkout man;

we were on first-name terms now. By "the usual," he meant pasta, pasta sauce, and cheese. Teaching myself to cook wasn't going spectacularly well.

"I'm all set for pasta at the moment."

Chortling at me, he scanned through my items.

I crossed the road back to my accommodation. Realizing that Toby would be having his meal around then, I paused for a moment and pondered what else I could do. I could go for a run, I thought. I hadn't ever been for a run, at least not voluntarily. I'd only run before at school, when I'd been forced to do cross-country first thing in the morning. And even with the best will in the world, that was never fun. I went into my room, put down my new purchases and my backpack, and got changed into my sports kit hand-me-downs from Mary. The leggings had holes in the crotch. Bit embarrassing, but I figured no one would be close enough to see them. Plus, I wanted to leave as quickly as possible. I could smell Toby's meal from my room, even. Christ, it was disgusting.

I set off for the meadows and did a couple of gentle laps. It was completely dark by that time, and the streetlamps glowed serenely, like flickering fireflies. I let clouds of fresh breath waft from my mouth, listening to my body's bumpy rhythm. It was quite nice really. I heard snippets of conversations as people strolled into town in groups, or tottered hurriedly with their earphones in. The mist hung around the tops of the trees.

I stopped in the middle of the grass. Stood still for a minute. Stared at the looming mound of Arthur's Seat.

I moved my eyesight down and watched the walkway through the middle of the meadows, which was lined with regal trees and tall, boxy lamps. A woman shuffled down the middle in a long trench coat. She was wearing a hat and had her dog on a red leash. I saw her observing the moonlit green grass, the trees, the lights. Even from a distance, I could make out the wrinkles around her eyes, her powdery skin, the creases on her lips. People dodged around her, their urgency and the racket of their loud voices disrupting her tranquil air. She seemed familiar, but unobtainable. I wondered if she had children. If she did, what was she like as a mother? Did she hug them and comfort them when they cried? Send them cards and gifts and poems in the post? Did she drive to visit them, wherever they were, if they needed her? She disappeared into the shadows, as if she'd never been there in the first place.

When I got back, Karen was cooking in the kitchen with all the windows open to let the smell of Toby's food drain away. I went in and poured myself a glass of water. She told me she'd just gone to a boring talk on careers in town planning, and we had a little laugh about the individual lengths we were going to in order to avoid Toby's cooking.

That night, I'd showered, done some work, and eaten pasta for dinner (the usual). I'd even lit the lavender candle I bought from Dave.

As I lay in bed, I'd flicked through Maya's photos on Facebook, my palms and fingers sweating, my heart beating fast. She'd added me the day I met her, but I'd made myself wait 'til the following day to accept. I hadn't looked through her profile yet; the thought had made me too anxious until then. It was almost as if I was scared that she'd know what I was doing or something, like somehow, she could see me.

I studied each part of her face in the photos, the shape of her body, what she wore. There were quite a few pictures of her with one lad, Ethan Giles. He seemed alright-looking. He was sort of tall, quite trim, and often had a cigarette tucked behind his ear.

In the photos he seemed to cling to her waist. She usually rested her hand on top of his shoulder. I wondered if he was her boyfriend but dismissed the thought quickly. There were loads of photos of her with other people as well, but I clicked on his profile anyway. He was also from London, studying at Cambridge. Around one a.m., I put down my phone and blew out the candle, watching the smoke curl up into the air like the train of a wedding dress.

A couple of days later, I stood there in that bar on Lothian Road, about to see her.

I felt a slight sense of fear in the midst of the rowdy drinkers, but I pushed through the sweaty cluster of bodies towards the bar. The lights were purple and stringy like zips, red and blue and yellow dots flashing on the ceiling. I could only imagine how confusing this would be if you were drunk. Horrible, in fact.

I couldn't see anyone I recognized. The music was unbearably loud, and I'd never heard these songs before. Something about people's butts and tits, drop it low, drop it hot, or some shite like that. I didn't enjoy anything about this place apart from the fact that Maya was in here somewhere.

At that moment, I saw Harry emerge from the toilets, doing up the top button of his jeans. He walked over to the far corner, and I followed his path.

There she was, I could see now, sat at the end of the red leather couch, along with all the others. There was a single lightbulb hanging above the table in front of them, which was sticky and speckled with rings from the bottoms of people's glasses.

As Harry sat back down opposite Maya, she turned and spotted me.

"Aisling!"

All I could feel was bubbling in my stomach and my legs moving mechanically. The noise and strangeness of the place seemed a distant memory.

Maya scooched over to make space for me to sit beside her. Her body relocated gracefully, even with it being such an awkward movement to perform. Generally, she seemed like she put no effort into looking beautiful. Her hair was like shining liquid, flowing straight down her back and over her chest. Her skin was glowing. Her black leather trousers and a sparkling top, embellished with jewels, hugged and hung off her body.

I took off my cross-body bag and my long black coat, stuffing them under the table onto the damp floor.

Maya raised her voice. "Everyone, this is Aisling."

I recognized most of them from the meeting the other day, and I dipped my head towards them.

"This is Harry," she said quietly, her jaw tense, "and have you met Isla and Isabel, maybe at the meeting? This is Rohan. And I've only just met you guys properly," she said, smiling towards the other two, "but this is Ruby and Tommy, who are also first-years."

Everyone greeted me; some of them even smiled.

Maya turned to face me.

"Do you wanna get a drink?"

"Oh, I don't drink," I responded quietly, moving my hair behind my ears. She turned her lips upwards. "But sure, I'll get a Diet Coke or something."

"You don't drink?" sneered Harry.

"Come on," Maya urged, shuffling me out of the booth and walking ahead of me to the bar. Once there, she turned on her heels to face me.

"Harry can be—"

"It's no bother." I shrugged, leaning on the bar. That surface was also sticky. "I'm used to it."

Maya raised her eyebrows. I could see her deciding not to pry, but I wished I hadn't said that. I wished I'd made some comment about how it was clear that Harry fancied her. I'd noticed it at the meeting on Thursday. He always blushed when Maya spoke to him, or spoke at all, in fact, and he couldn't stop staring at her across the room.

Looking behind the bar, Maya attempted to make eye contact with one of the bartenders. As we stood there, she wiggled her body along to the music. I so admired her confidence and freedom. She had everything sorted; she was a solid and complete human. The way she could chat to people with such ease and make them feel comfortable. I wished I could have that effect on people.

A woman behind the bar came over to us.

"Yup?" she mouthed through the gum-chews.

"I'll have a white wine, please. Ash, you?"

"Oh, I'll just grab a Diet Coke, cheers."

Maya paid. I insisted I'd get the next one. As she told me not to worry, she brushed her hand on mine. Her skin only touched me briefly, but I could feel its softness.

"So then," she inquired, leaning over to me, "no alcohol? I respect that. Every morning after I've had something to drink, I tell myself I'm going teetotal."

"Sure, I can imagine," I laughed. "I guess I've just never drunk really, my mother—"

What was I doing?

I'd gotten carried away. Maya did something to me. I felt relaxed and uninhibited, but also mad and sweaty and juddery. I never spoke about my family, and if I did, she'd probably think I was mental.

"She just really . . ." I continued, grasping for words.

I could feel Maya looking at me intensely, furrowing her brow.

"Really hates it. Hates the drinking."

By the way she was looking at me, I felt like Maya could see through my eyeballs and into my brain. As if she could read the words which were on the tip of my tongue, the ones I was trying so hard to hide: My mother is an abusive alcoholic and I'm scared of turning into her.

Maya ran her gaze over my face like a damp, refreshing cloth.

The barwoman smacked two glasses on the counter.

"Thanks," Maya said without turning around.

She picked up my Diet Coke and handed it to me. As I took it from her, I tucked my hair behind my ear with the other hand.

Gently, she caught my wrist in midair as it retreated from my head. Our skin contact made my muscles weaken, starting from my chest, spreading all the way down my belly and legs. She held my hand out in front of her, turning it so my palm faced upwards. We both examined the scar there, and in one motion our eyes met again. Our hands were still touching, and my body felt as if it were getting weaker, weaker, but beating stronger and stronger.

"That scar."

I tucked the hand into my pocket, pulling away from her touch.

"Sorry, I shouldn't have—that was—"

"No, you're fine." I blushed. "It's from way back."

"It's not a stigmata or something crazy like that then?" She laughed uncomfortably.

I shook my head and felt my face going even redder. I could faintly smell the wine on her breath as she exhaled.

"Your flatmate," I said, changing the subject, "you know him from your course then?"

"Gabe?" she said, swallowing a sip of wine. "No, actually. We met through a mutual friend from London."

"Londoner?"

I acted like I hadn't already known this fact about her.

"West London. Why, do I not seem like a Londoner?"

"Oh, no, you do alright. You've got that sort of air about you."

"What sort of air?"

As she tipped her head to one side, I surrendered my free hand, my lips tight together.

"I dunno. That sort of confidence about you," I laughed.

She smirked and then looked down at my shoes. They were old black boots that my mother had gotten me in a charity shop in Clare. Tatty and flimsy, nothing like the pristine ones she was wearing.

"Confidence." She tittered. "It's an act really."

Her voice was quiet, and as she peered back at me, I flicked my gaze between her eyes. I knew what that felt like, to have a protective exterior.

After a moment, she continued. "Do you want to come over to our flat sometime for dinner?"

I took a sip of my Diet Coke, holding this eye contact with her. I felt my body gaining strength, the nerves and twitchiness deflating.

"I'd like that."

Maya nodded. "Good."

That evening, I chatted to more people than I had done since I arrived in Edinburgh. What surprised me most was that I had things to talk to them about: poetry, books, lectures, the city. I spoke to Tommy, Ruby, Isla, but mainly, I spoke to Maya. I knew where her body was in the room at all times, even though I desperately tried not to look in her direction too often. I could feel where she was sat or stood. I listened for her voice and her words, trying to grasp onto any part of her that I could.

Every time Maya finished her drink, she would catch my eye and we would go up to the bar together. I felt as if I were changing, morphing, even over the course of just those few hours. It was beginning to feel as if nothing before that had ever existed.

12

DINNER

Maya

I STOOD IN the kitchen, chopping vegetables. There was music playing on the speakers, and every few minutes I stopped to take a sip of wine. I'd spent ages thinking about what to cook for Aisling while I'd been sat in the library that day. All the ingredients for dinner were piled up on the counter, the colors vibrant against the blue tiles on the wall.

That afternoon, once I'd finished studying, I'd run to the shops, then showered and changed into a long skirt and a white T-shirt before spraying perfume on my neck and wrists. I had put on some makeup, but not as much as normal. Just mascara. My hair was drying naturally into curls. I'd washed it without thinking, and then realized I didn't have time to dry and straighten it.

I felt pretty anxious about Aisling seeing me like this, but it was too late for me to do anything. She was just about to arrive.

I had been all over the place lately. I felt distracted and my sense of timing was off, which was unlike me. I constantly seemed to be late for things, and chunks of the day just disappeared.

There was a knock at the door, and I hurried through the corridor, stopping to check in the bathroom mirror that my face looked OK. I caught a glimpse of my hair, which had sprung out into frizzy ringlets. I shut my eyes for a moment and let my chest sink.

When I answered the door, Aisling was stood there in her black coat, clasping a bottle of white wine, a grin spread across her face. Her eyelashes looked darker than normal, and her lips were the shade of poppies. I could smell her sweetness, like fresh air and lavender.

As soon as I set eyes on her, the chaotic feeling in my stomach faded. In its place, I felt my breath unwind and become languid.

"I tailgated someone into your building."

"Excellent," I said, echoing her grin.

"For you." She held out the bottle.

I thanked her and took it, quickly beckoning her into the flat so she could get warm. The hallway was always freezing.

"I remembered you drank white wine at the bar the other night."

I shut the door and examined the bottle. "It's perfect."

She was looking at me, at my hair, my cheeks, my eyes. After a pause, her hushed voice washed over me.

"You look"—she paused, swallowed—"beautiful."

My throat weakened. Starting to remove her coat, she stepped towards me, shuffling it off her shoulders and pulling at the sleeves. When our eyes reconnected, she was near to me. I wasn't sure how close we were to each other, but it was close enough that I could feel her breath on my skin. I leant over to take the coat from her arm.

"Thanks," she whispered.

The smell of toothpaste. The warmth of her soft breath.

I shut my eyes. Our faces, our bodies, were only narrowly parted.

At poetry society the Thursday before, Aisling had read out something she'd written.

"I wrote this a while ago," she'd said, her eyes briefly, almost imperceptibly, glancing towards me, "so it's not recent. But if you'll bear with me, this is new for me, reading my stuff out."

The poem had been about a girl. It described the way their two bodies touched and moved together. The tension of their lust with the feeling of secrecy. It drew on imagery from the church: stained glass windows compared to the stained bedsheets, the crucifix likened to the shape her body made on the bed as she lay there, quivering, coming.

While Aisling read it, I had closed my eyes and listened to the way her voice moved over the words. Something had erupted inside me. My eyes had swollen with tears, which trickled down my face. My eyelids felt hot and tender to touch; my throat ached.

After she'd finished, the room fell silent. It floated in space. I got the feeling no one really wanted to read any of their own work after that. Perhaps they felt it seemed limp in comparison.

When I got home from the meeting that night, I'd spoken to Ethan on the phone. He'd told me about his day, about how his work was going, when he was thinking of heading back to Cambridge. He said how much he missed me. When he asked how things were going in Edinburgh, I didn't feel like talking about it. Ethan wasn't within my orbit, and I was well outside of his. I was becoming more aware of it. But I reminded myself that this was Ethan—Ethan, who I'd always wanted to be with, who everyone thought I should be with—and I tried to push any other thoughts from my mind.

After we hung up, though, I lay there thinking about Aisling. I shut my eyes and reminisced about the way her mouth moved when she spoke, about the way her face crinkled when she smiled. I reflected on how easy it felt to talk to her; our conversations like a constant flow of clear water. I thought about her mind. The way she chose words which no one else would think of. As an extension of her, her writing made me feel something I never had before.

We stood there in my flat. I cleared my throat. I used all the strength I had to open my eyes and move away from her. I hung up her coat. She moved away from me, too, acknowledging that moment for a second, only subtly, before gliding further down the corridor into the flat.

"Grand place you've got here."

"Gabe found it really," I admitted.

"He's not around?"

I shook my head. Aisling peeked into the kitchen, her hands resting on the door frame before speaking again.

"The food smells delicious. Can I do anything at all?"

"No, no, not at all, thanks. It's almost done," I responded.

I marched towards her briskly, brushing against her softly as I entered the kitchen.

"I'm honestly in awe of you being able to cook," she guffawed. "I can do pasta and pasta alone. No, honestly, I wish I were lying. I never learnt, I wanna teach myself or—"

"I can teach you?"

"Yeah?"

"I mean, I'm not great, as you will see"—I giggled and paused to touch my hair—"but my parents, well, my dad showed me the basics."

I faced the stove, turned up the heat, and stirred the pot as it bubbled.

"What about your parents?" I asked.

Aisling's silence carved into the air. I almost twisted back to look at her, but instead, I just listened. I could feel her moving her feet, taking a big breath in, retreating to the wall on the other side of the room.

"They always, sort of, told me what to do, but they weren't exactly great at teaching me much."

I turned around and let my hands rest on the counter behind me. "In general?" I inquired.

Aisling swallowed and flared her nostrils, scrunching her mouth into a small round shape. "They're not the most accepting people."

We breathed and watched each other. I felt like I'd known this about Aisling ever since she pulled her hand away from me when I saw that scar on her palm.

"Your poem," I muttered.

Her face didn't move, but her bright eyes glanced between mine. Smiling faintly, she nodded. "I'm not with her anymore, by the way. The girl from the poem."

I couldn't think straight or capture my emotions. My insides twisted and pulled.

I knew that at some point I had to mention Ethan, but every time the thought of him went through my head, I didn't want to tell her.

I moved back to the stove, saying the food was just about ready. After serving up, we took our plates through to the sitting room, where there was a bigger table. I went back to fetch water, glasses, and the speakers. As I reentered the room, she heard the music. It was a Nick Drake song. She said she loved Nick Drake, and I could only smile at her. I loved him, too. I asked her who else she liked. She told me she'd class her taste as assorted. She was into a bunch of different music; she loved Van Morrison,

T. Rex, AC/DC, Bob Dylan, and, of course, Nick Drake. She reciprocated the question, and I told her I thought Ella Fitzgerald, Aretha, and Beyoncé comprised the soundtrack to my life, but I loved Nick Drake, too.

Aisling grinned and thanked me for the meal before we started. We ate and chatted about her lecturers, laughed as we recollected funny things about the bar night a couple of weeks ago, and gossiped about various things we'd read recently. She asked me all about home, and about my family and friends. Over the course of the meal, I felt myself relaxing. It was so easy to be around Aisling, effortless talking to her. I didn't have to think or pretend.

Once we were finished, she leaned forward, resting her elbows on the table.

"Maya, can I ask you something?"

"Always," I jested, sitting back in my chair, linking my fingers together.

"How come you invited me over?"

I swallowed. "What?"

"As in, you've been so good to me since I met you, I wondered why?"

I felt myself getting hot, warm blood trickling into my hands and chest and cheeks. My heart creeping up into my throat. I tried to conjure up what to say. I tried to align my thoughts with my words, to weave together my feelings with my outward persona.

"Aisling—"

The buzzer for the flat went off. Three sharp beeps.

It was as if someone had just awoken me from a dream that I didn't want to end, and I was grasping at the air to retrieve it. I was linked with her, unable to get up or move, unable to speak or articulate my thoughts, unable to create clear thoughts, even. I waited and moved my mouth around, but nothing came out.

The buzzer went off again.

"One second, OK? Just one second."

I walked into the hallway, pressed down the receiver.

"Hello?"

"Maya!" said a voice. "Buzz me up!"

I waited with my finger still pressing down. The insect in my body awakening from its slumber and flitting through my bloodstream. My breathing halted. My mouth filled with saliva as my brain slowly recognized the voice.

13

LEAVING

Aisling

I GOT OUT of Maya's building quickly. The door shut behind me, and I stood with my back against it for a moment. My coat hung down below my knees, and rain was spitting from the sky in heavy droplets.

I breathed. In through my nose. Out through my mouth. I shut my eyes. They were wet. I pressed my eyeballs into my head with the palms of my hands.

Stop, Aisling. Stop.

I picked myself up and hurried down the stone steps, away from her and away from her building. I crossed my arms and tucked my head down. After rubbing my eyes again, I looked at my hands. They were black with mascara and eyeliner. The inky liquid clambering around the wrinkles of the small scar on my palm. This was the second time I'd worn eye makeup and the second time I wished I had never put it on. It had been a ridiculous idea to wear it.

Why was I so upset? Perhaps I liked Maya more than I'd realized.

But of course she had a boyfriend. Of course she did.

I kept walking fast. Head down, long strides.

I'd known it was coming, but I'd been in denial. I had been able to tell from the photos that she was dating that lad Ethan, but I didn't want to believe it. I wanted things to be different.

As I walked, I cringed at myself. The wine, the question about why she'd invited me, telling her I wasn't with Orla anymore, getting close to her in the hallway. It'd all been so stupid, and I hated myself for it.

At the last poetry society meeting, I had decided to read my writing about Orla. I'd drafted it over the summer. It was about my family as well, contrasting the way I felt around them with how I felt around her. I didn't feel ready to read something out at society. In fact, I felt so scared about sharing it that I shook as I spoke. Putting my words out there was terrifying, especially with something so deeply personal. It was like handing over pieces of my life, the deepest parts of my thoughts and feelings, to people who were mostly complete strangers.

The sole reason I'd done it was to communicate to Maya where I stood. I wanted her to know. But in the end, I discovered that by reading it out, I felt liberated. I felt accepted through the solidarity of people listening to me and sharing in those words.

After the meeting, Maya had messaged me.

> Your poem was incredible Ash. Do you
> want to do that dinner on Saturday? X

I'd messaged back the following day telling her I'd love to. She'd told me to arrive around seven p.m. and had given me her address.

Early on the Friday morning, I'd bumped into Gabe at the same coffee shop as before. I made a thing of going in there just in case I saw Maya. As I entered, he'd been standing in the queue. I'd caught his eye.

"Gabe, right?"

He'd taken out his earphones. "Yes, you're Maya's friend, aren't you. Aisling?"

"That's me."

We'd shaken hands with two short movements as if this were a business exchange, which made me laugh inside.

"I think I'm coming for dinner on Saturday, will you be around?" I'd asked him.

"Oh, no, sadly not. In fact, Maya asked me when I'd be out before she arranged it with you," he'd laughed.

I blushed a bit and my breath caught in my chest. I knew she wouldn't want me to know that, and I could tell Gabe immediately regretted mentioning it, but the idea of it made my heart race.

"Next!" the man behind the counter had called.

Gabe had asked him for a mocha with extra foam. Pushing his glasses up his nose, he'd turned to me and said, "Hopefully we can all do something another time, though?"

I'd nodded enthusiastically before he sheepishly left the café.

I'd walked home, coffee in hand. When I got back to my room, I sat at my desk chair, swinging side to side, looking out at the trees. I had some work to do, but I couldn't focus.

I'd pulled my curtains together, taken off all my clothes, and wrapped my towel around my body before walking down the corridor to the shower room. Turning it on to the coldest setting, I'd analyzed my flat chest and pale face in the mirror before stepping in. The water froze and shook my skin. I let out small shrieks as it shocked and aroused my body; then I turned the temperature warmer and touched myself until I came. After scrubbing my face and my hair, I got out, went back to my room, and sat in my soaking-wet towel. My teeth jittered as my body relaxed, and I lay back on the bed.

Saturday arrived. I went for a run before getting ready for dinner. I'd made a routine of this, mainly to avoid Toby, but I also enjoyed stopping at the same place in the meadows and watching the people walk down the pathways. I'd noticed the running in my body, too. I felt more alive, happier, recently, and my shape was changing. My legs were becoming more muscular, and so were my arms. After getting back from my run and washing, I dried my hair, lit my candle, and sat in front of my mirror, trying to put on the makeup which I'd bought from Dave. My eyes looked darker, and this time I'd managed to do both of them, which was an achievement.

On the way to Maya's house, I'd picked up a bottle of wine for her from the shop. I felt weird about getting it. I'd never bought alcohol before, but I knew she'd like it and that was all I cared about. When I arrived at the flat, I felt sick to the bottom of my stomach. But I knew that just being around her would make me feel calm.

And then, after dinner, that buzzer had gone off.

It'd all happened so quickly. Maya came back into the room to tell me her boyfriend was here to surprise her, her face expressionless. I'd got up from my chair immediately.

"Aisling, I'm so sorry," she had pleaded, looking at me, trying to pass me a message with her eyes. It reminded me of the time I'd tried to do just that with Orla when Sean had caught us together. "I had no idea," she continued.

"It's honestly fine, sure, no problem!"

Just then, Ethan had walked in. He was wearing a puffer jacket and had a cigarette tucked behind his ear, just like in all the photos. A gust of outside air followed him, and a sheen of sweat coated his skin. I wanted to leave right away.

"I'm Ethan," he said, raising his hand to me.

He put his arm around Maya's waist, which made my chest tighten. I looked at her; she didn't move.

"Aisling," I said, forcing a smile. "How was the journey up?"

"Long, but worth it."

Maya met his eye and smiled without showing her teeth.

"Have you changed your hair?" Ethan muttered, touching it with the tips of his fingers.

"Oh." Maya shook her head. "I completely forgot."

"Well, I best leave you alone. Thanks so much for dinner, Maya, it was great. Let me know how much I owe you."

I walked past her and out into the corridor. My head spun. And with that I left, the door slamming shut behind me.

14

TWO WORLDS

Maya

Cramond Beach, Edinburgh

ETHAN AND I walked silently across Cramond Beach. He held a cigarette in one hand, between his thumb and forefinger, and his other hand clasped one of mine. His skin felt warm and dry. Smoke blew out from the corner of his mouth.

The tide had retreated, revealing the camel-colored sand, leaving patches of seal-gray and silver water which rolled like wrinkling skin. Coal-like rocks, covered with moss, were lumped around the outskirts of the beach, and out beyond, the sulky clouds reflected their own glum expressions. We were wrapped in warm coats. I had on a red scarf and a fluffy black hat, and Ethan wore a cap. I felt almost back to normal, polished, preened, and made-up.

Once we reached the causeway, we looked out to Cramond Island. The concrete of the walkway towards the island was lined with spiky stones which protruded upwards like misshapen Egyptian pyramids. Ethan stubbed out his cigarette on one of them and blew the last bit of smoke from his lips.

"Shall we go across?" He squeezed my hand tight. I nodded.

We walked across the causeway, holding on to each other.

I thought about the night before, about how I'd felt when I'd heard Ethan's voice over the buzzer. He'd looked so fresh-faced from being out in the cold. Pale but rosy, his dimples pressing deeply into his cheeks, his bag slung over his shoulder.

Ethan meeting Aisling felt like two worlds colliding. She was part of a secret place that I was only just discovering. Around her, I removed my exoskeleton and lay bare. Ethan brought me back to reality, back to the place I'd been since I was a little girl. There, people saw me as the person I presented, the person I had spent so long fashioning.

Aisling had exited so quickly when he arrived, crumbling away into the mist. After she left, I went to the front door of the flat and waited there, listening to her footsteps echo through the hallway. The weight of the door to the building slamming her out, dividing us.

When I came back inside, Ethan had opened the bottle of wine which she had brought.

"Aisling bought that," I snapped.

"Was I meant to leave it unopened?" he'd said, taking a sip straight from the bottle. "Weren't you guys gonna drink it anyway?"

I began to tidy, stacking up the empty plates and water glasses, turning off the music.

"She doesn't drink, she just bought it for me."

"Doesn't drink!" he'd scoffed. "Mental."

I took the plates through to the kitchen. My brain felt like a fuzzy, static TV screen. I knew there was a reason Aisling didn't drink, one I didn't understand properly, and that Ethan understood even less. I sat there while Ethan ate the leftover dinner and drank the wine Aisling had bought.

I knew I should be grateful that he'd just made the journey to see me. I knew I should be overjoyed to see him.

We had sex that night, but I kept my eyes closed most of the time and didn't move much. I felt anxiety seeping through my body like honey running off a spoon. I attempted to keep images of her at bay, but as I lay there, I couldn't prevent my thoughts from moving towards Aisling. My mind spun like a swinging wind vane, pointing one way, then the other.

In the morning, I suggested we should drive out to the beach in Ethan's car. Perhaps we needed to get some fresh air, I told myself. Space to clear my head and come back to center. The insect nibbled at my stomach lining, but I forced myself up and got myself together. Makeup on, hair styled, clothes thoughtfully chosen.

As Ethan and I walked across the causeway, the wind blew us from all angles. It tasted salty and thin. We could see boats in the distance, blue and white dots of wood and sails, rocking like lullabies. When we got to the island, we walked to the small beach.

"Shall we sit?"

Ethan nodded. We perched ourselves down on the sand and faced back towards the mainland. Water lapped in front of us, foamy and pigeon gray.

"Maya, what's going on?"

"What do you mean?" I laughed.

"You don't seem happy to see me."

He almost sniggered as he said this, as if in disbelief. Flicking some sand with one of his fingers, he looked down to the ground between his legs. In that moment, I saw that little boy who I'd always known and loved. I'd yearned for this for so long. What was I doing? My heart sank down into my stomach. I put my hand on his leg.

"I am. I am happy to see you."

We waited.

"I'm sorry, I just feel so far away from you," I said.

I put my head on top of his kneecap, clinging to his leg with my arms. Looking back to the land from this island, I thought about how, from a different perspective, it looked so cartoon-like and constructed. Ethan's hand rubbed my back tenderly.

"I thought so. That's why I came to visit. I can do that more if you'd like. I'm happy to."

I closed my eyes. "But it's so far."

After a moment, he took his hand off my back.

"Maya, do you want this?" His voice was louder. The water lapped more violently on the shore, and the wind picked up. I sat up to look at him. "Do you want us to be together? Because recently it seems that you don't. We're barely speaking, and you haven't even mentioned about when you could come visit me."

I felt panic rising into the tip of my head. I was a hot-air balloon, fire blowing within my body, lifting me towards the sky. It was true, I hadn't suggested a time to go and visit. I hadn't been a good girlfriend. I was being selfish and inadequate.

I thought about what everyone would say if we broke up. I imagined Naomi and Tim, who I texted constantly and spoke to on the phone most weeks. They'd tell me how sad they were and ask me how we were going to navigate this as a friendship group. I considered my parents saying how terrible it was that we'd broken up. I reflected upon their secret belief that I would never find anyone as good as Ethan. I thought about all my friends from home, Gabe, people from poetry society. What would they say? I felt sick at the idea of Ethan being with other people and at the thought of how upset he would be if our relationship, something we'd both seemed to want for so long, failed.

This was me. This was what I had always wanted.

"No, I do," I assured him. "I want this so badly. I'll be better, I promise. I'm sorry. I'm so sorry."

I looked at his familiar face, his eyes, his wind-brushed skin.

"I'm happy you're here, thank you for making the journey for me," I whispered, tears brimming in my eyes.

15

RUBY'S

Maya

Edinburgh
OCTOBER 2013

IT'D BEEN WEEKS since Aisling had come to my house for dinner.

Having resolved to focus on my relationship with Ethan, I'd waited to message her until after he had gone back to Cambridge.

> I'm sorry again about the other night. It was lovely to have you for dinner—we should do it again sometime. X

To my surprise, Aisling had responded the following day.

> No problem—thanks for having me and hope you two had a nice time. I'll see you on Thursday.

Sure enough, I'd seen her at poetry society that Thursday. She arrived late and left early, so I didn't get the chance to chat to her. I did wave across the room, though, and she had smiled back at me, lifting the palm of her hand, the ball of scar tissue bundled in the middle of it.

Every week of October, I made a special effort to look nice on Thursdays, wearing makeup and really considering which clothes to wear. I always looked forward to seeing her, especially since I'd figured everything out with Ethan. I assumed Aisling and I could just get on with being friends.

But every Thursday, despite my efforts, the same thing happened. She would arrive to the meeting late, we'd acknowledge each other, then she'd leave before I could catch her.

One week, we had a quick exchange at the end of a meeting.

"How are you doing?" I'd asked. "All good?"

She'd responded, "Yeah, all good, thanks, that poem you read out was class."

I felt all the muscles in my stomach contract.

"Look, I've got to run, but have a good evening," she'd said, walking away.

That was it.

In the middle of October, I'd been down to visit Ethan in Cambridge for a couple of nights. I had a great time with him, but Aisling was constantly at the back of my mind. All I wanted was to chat about why things were so strange between us. I wanted to spend time with her and straighten things out.

The last meeting of the month fell on Halloween. At that meeting, Ruby, a first-year, announced that she was having a party that weekend.

"I'll put my address in the group chat," said Ruby, standing up at the end of the meeting as people were packing up their bags. "It's on Saturday. You're all welcome to come, bring whoever you want."

I messaged Aisling after that meeting, the first time I'd done so in weeks.

You coming to Ruby's at the weekend? X

The next day, she responded, saying she was thinking about it. I replied, saying I hoped to see her there.

The evening of the party, I showered after I got back from the library, got dressed, had dinner, and headed to Ruby's flat. I had invited Gabe to come with me, but he was busy.

Ruby's place had flaky walls where the paint was peeling and it smelt faintly of weed. Music pumped loudly from the speakers as I said hello to everyone. There were maybe fifteen or twenty people there, and a handful of them were from poetry society. I scanned the room carefully, looking at the people hanging out the open windows, smoking, or lounging on the furniture. I couldn't see Aisling anywhere. I grabbed a drink and started mingling.

About half an hour later, Aisling walked in. Her long black coat on as usual, and her hair draped over her shoulders, shining like a dark mirror. Ruby hugged her and welcomed her in. I caught Aisling's eye and she smiled at me before going to greet some of the others. I took a sharp breath in and continued to chat to people, keeping my peripheral vision on where Aisling was standing.

After a while, I went over to the drinks table to refill my glass. I knew Aisling was in the kitchen with Ruby. I hoped that if I stood there, I could catch her when she came out.

"Maya," I heard. I felt Aisling's presence before I saw her. She walked out of the kitchen and came to stand with me at the table.

I turned to face her, feeling suddenly lost for words. I said hi and exhaled, my stomach going tender. Finally we could get everything resolved.

"It's been ages," I muttered, brushing my hair out of my face, readjusting it, then crossing my arms.

Aisling reached for the orange juice and tonic water, concocting a fizzy tangerine-colored mixture in a glass. There were so many things I had wanted to talk to her about these past weeks, but my mind seemed to go blank now that she was right in front of me.

"How've you been?" I asked.

A sigh filled her chest, and then her shoulders sank as she released the air. She looked down into her glass before speaking.

"It's been a while. I'm sorry about that. I've had— It's been a mad few weeks, settling in and all."

Her head tipped up and we looked into each other's eyes. A shiver ran across my shoulders and arms. I hadn't had this feeling in weeks.

"Sorry—" she whispered to me, as if we were the only two people in the room. She gave me such an intimate look. I knew then that we had missed each other, that it wasn't just me feeling these things.

My arms collapsed in front of my body. I wanted to take her in my arms, to hold her, but instead I just put my palms on my chest.

"Aisling, please don't say sorry, I should be apologizing. I should have—"

She stopped me from speaking with a small gesture, one which I knew most other people wouldn't even have seen her perform. It was the slightest flick of her head and fingers.

I steadied.

"Friends?" I asked.

Aisling sipped her drink. "Friends," she confirmed.

It felt as if, in that moment, we had both acknowledged the strangeness of the last few weeks. We acknowledged it and let it fade into the past and dissolve. I wanted her in my life as my mate, and I knew that she wanted that, too.

16

LIFE DRAWING

Aisling

DECEMBER 2013

MAYA AND I laughed and chatted away as we took off our backpacks and tucked them under the benches. Coats and scarves came off, too, and we hung them up. My cheeks burned from being out in the bitter cold, and our hair glistened from the moisture in the air outside. Maya patted hers, feeling the wet and shaking it like a dog. It had started to go curly from the damp. I loved it when she had her hair that way.

We took two of the paint-splattered aprons off the hooks and popped them over our heads, helping each other tie them up at the back.

"The life drawing session will commence in five minutes," hummed a woman wearing a pair of green-winged cat-eye glasses and a loose black jumper with flared sleeves. "Please come and set up at your easel."

It was the middle weekend of December. Our exams had happened straight after the revision period, so we'd finished them a couple of days ago. In a week or so, all the other students would be done, too, and everyone would be heading home for the holidays.

It'd been a couple of months since I'd been for dinner at Maya's apartment. I had avoided her for weeks after that evening. I felt embarrassed. I'd assumed that the spark between us had been something more than friendship, but when Ethan arrived, I realized I'd obviously been wrong. My embarrassment and sense of self-preservation meant that over those weeks I sort of hid myself away a bit. I mostly got on with my work, went for runs, did some writing, and read a lot.

But the thing was, after that time of barely speaking to her, I felt like I had a hole in my gut. I missed talking to her.

When we had been together those few times, and when I was here in Edinburgh, I could almost forget about my past. I'd been released from all that, like I finally had managed to mend a wing and could fly above houses and cities and fields and see everything from a different angle. Like I could be fun. And joyful. I wasn't constantly filled with dread, with flashes of darkness or cruelty. She was a part of that change.

As I lay in bed one night in late October, I decided that I wanted her in my life, even if our relationship would never be romantic. Sure, I wanted something to happen with her, but if it wasn't going to (which it clearly wasn't, as she had a boyfriend), then I wanted her there as a mate.

When I'd seen her at Ruby's party, I felt as if we both knew what was going on. There was an unspoken understanding of what had happened, and because of that, we could wipe the slate clean.

For the last month or so, since Ruby's, Maya and I had been hanging out almost every day. Going for coffee, walking to or back from poetry society together, sitting in the library next to each other working or revising for our exams. She'd even taken me to the bookshop she loved, just past Grassmarket. As we entered, it was almost as if she was unveiling a secret place, a special part of her.

"I could live in here. Right here, in this bookshop," I'd said to Maya as we walked around it and scanned the walls, which were jam-packed with books of all colors, shapes, and sizes.

"Shall I ask if they rent it to students?" she said, a glimmer in her eyes.

I chuckled as I ran my hand across the bumpy leather spines which stuck out from the skew-whiff walls. We'd both chosen and bought books for each other that day. She gave me *Sister Outsider* by Audre Lorde, and I gave her a collection of Emily Dickinson poems.

I placed *Sister Outsider* on my bedside table; it was too precious for a shelf. Before I went to sleep each night, I would read each line delicately, like it was a ripe berry I'd tweaked from a bush. I'd pop it in my mouth and sit there, sucking at it, enjoying the sweet, powerful taste. My favorite essay was "Poetry Is Not a Luxury." Poetry is a necessity, wrote Lorde; it is light and illumination, something which names and forms things which otherwise couldn't be captured.

One weekend in November, Maya and I had gone on a day trip, just the two of us. Gabe had his car with him in Edinburgh, and Maya said she could go on the insurance for the day. She'd messaged me:

> I've always wanted to go to Cardrona Forest. Takes like an hour to drive . . . we'd have to leave early so we can catch the light, but could be fun, maybe at the weekend? X

I'd messaged back telling her I was keen. That evening, I'd ordered some secondhand walking boots on eBay for next-day delivery, and that Saturday, Maya had picked me up early in Gabe's car. As I trundled down the steps of my building, in my new (old) walking boots, yawning, with my backpack full of food and water, Maya had rolled down the window and grinned at me so widely it made my stomach tumble over itself like crashing waves rolling onto the shore.

We drove for about an hour and fifteen to get to the forest, playing our favorite music, telling each other what we liked about it, and sharing memories about why certain songs were sentimental to us. I looked at the colors of the Scottish countryside as we drove. It was stunning in a different way from Ireland. The bumpy hills loomed in the distance, mulberry purples soaking into bright greens, which seemed dampened in the winter fogginess. Pine trees and fir trees and spruce trees stood tall above us. The soggy weather made them droop like wet seaweed. As we got further into the countryside, the air became thick with water vapor droplets, which peppered the windscreen.

At the forest, we parked up and wrapped ourselves in our waterproofs. It wasn't raining as such, but it was one of those days where the sky sprayed a mist which sticks to your skin and clothes. We couldn't stop giggling at how ridiculous we looked, our hoods tight around our faces and our backpacks high on our shoulders. The sound of Maya's laugh was like boiling treacle. It just bubbled gloriously through the air, full and rich.

As we strode through the magical forest, trunks ran up into the air like veins, desperate to breathe. The mist hung like a layer of feathers, hugging the tops of the trees. The colors: figgy, fiery, sandy, mossy shades melding and melting into one another. I don't know if it was the weather, or Scotland, or me, but I never breathed so deeply in Ireland as I did then. All my senses woke up when I was with Maya. I had a whole new paint palette. The smell of the forest, that scent of pine and fresh, damp wood like nutty smoke or burning root ginger.

While we walked, I felt we were the only two people in the world. Our conversations were lighthearted, we joked about things we'd seen on the news or embarrassing stories from primary school, but each exchange held more and more meaning. The truth was, the more I got to know Maya, and the more time I spent with her, the more I liked her. She was so kind and intelligent, but she also had this warmth and generosity which enabled her to make everyone around her feel at ease. At the same time, I sensed her remarks of self-deprecation, which were not uncommon, reflected some deeper feelings she had inside her. It made me want to tell her I wished she knew how wonderful, how beautiful she was.

At the top of a hill, we'd eaten our lunch and looked out over the view.

On the way back to the car, we ambled even slower than we had on the way up. I didn't want that day to end. Even the drive back felt like it was going too quickly. As we sat there like drowned rats, with the heating on full blast, I kept my eyes on the road ahead. The sky dimmed, and the lights from the cars made the rain-stained road seem like a copper or silver liquid. Sometimes when Maya was talking, I would turn to her. Her ringlets curled, winding gorgeous pathways down over her shoulders and chest.

After she dropped me home that night, I had a hot shower and sat in bed. I just sat there, recalling everything that had happened that day. When I was with her, I didn't have to try to be anything. I was Aisling, and that was all.

In early December, Maya had seen an advertisement for life drawing classes.

Do you fancy life drawing class next weekend after our exams are over? I've always wanted to try x

I'd been in my kitchen, cooking pasta (again), when she sent me the message. My body had shivered when I saw her name come up on my phone screen.

I knew that Maya's mum was an artist, and that she had wanted to apply for art college after school, so I was guessing she was going to be good at drawing. I, however, would be horrible. It sounded fun, though, sort of, so I replied:

Sounds grand, but just a warning: I'm terrible at art. Might look a bit daft next to you (yes, I've seen your doodles in poetry society).

She'd responded instantly.

Don't be ridiculous. We're going! I'll book us in. X

And so, we walked into the studio. Aprons on. Easels set up.

It smelled of fresh paint, mucky water pots, and half-dried paintbrushes. The walls were the color of dirty linen, and the floor white oak, speckled with blotches of paint. Shades of lavender, denim, salmon, butter. Inadvertent blotches. Accidental plops. Concealed dashes.

As we set up next to each other, Maya picked up her charcoal and looked at it in her hand. Then she started to draw on the paper in front of her. Using the different sides of it, she made lines which were soft and wide, ones which were sharp and short, just trying things out in the corner of her page.

"Jesus. This is gonna be more embarrassing for me than I thought," I said.

Maya smiled, tilting her head to look at the paper.

"You've done this before then, I'm guessing?" I added.

"I did some drawings last summer, but not many." Curving the charcoal round like an extension of her hand, Maya lifted it off and pressed it onto the page.

"What kind of things did you draw?"

"Oh, just people, things like that."

"OK, everyone. Time to begin," said the woman with the winged glasses.

At the center of the room, she stood next to a block covered by a garnet sheet. The woman explained to us that the model's name was Leila. Leila would come and sit on the block, changing positions whenever it was easy for her, so we could do multiple drawings of her body. The woman said we could begin once Leila was settled.

Leila came out from behind a curtain. She was stunning. The way her figure sashayed across the room. I saw it shot by shot, like pressing a shutter button on a camera. She draped her body over the block, and we started to draw. I tried to capture her beauty. Her shape, her hair, her piercings. I looked at Leila's legs and saw a small scar on her knee. Her scar held a story, sure enough, and I wondered what it was.

I looked over at Maya's drawing after a while.

"Christ! That's unreal," I told her.

Her drawing looked like a real person moving on the page. She'd managed to gauge the proportions, the shadows, the light, all of it, flawlessly. Maya scratched her nose and analyzed her paper.

"It's not," she cooed.

"Well, if that's your attitude, don't be looking at mine."

I tried to cover my own drawing with my body, but Maya gravitated towards me.

"Aisling—"

"I know, it's bad, but—"

"It's perfect," she said. "It's yours."

I laughed a bit. I could feel the warmth of her body near me.

After the session was over, we said thank you to Leila. She got dressed and came around to see our pictures. I told her I was sorry; it was my first attempt at drawing. She looked much, much better in real life, I insisted, a sentiment which she seemed to appreciate.

We went to collect our bags and coats. Having sprayed the pictures to stop them from smudging, we rolled them up and popped them in our bags. Maya and I giggled to each other as we left the studio, already planning when we could come back and do it again.

Outside on the pavement, under a streetlamp, we stood facing each other. The light created shadows on Maya's skin, outlining the shape of her features perfectly.

I thought about what it would feel like to draw her.

"You've got some charcoal on your nose," she breathed, leaning forward to rub it off. The feeling of her fingers on my skin. The smell of floral perfume from her wrist.

The action reminded me of when I'd wiped that bit of ketchup off Orla's nose all that time ago. I wanted it to end the same way. I wanted Maya to lean in and kiss me.

I looked into Maya's eyes, one by one. She slowly withdrew.

"So." I looked down at my feet briefly. "I'll see you Thursday for the last meeting of term? Or maybe before, we could—"

"I can't," she said, tucking her hands into her coat pockets.

"I'm going to Cambridge tomorrow. I'll be traveling back here on Thursday, but I don't think I'll get back in time for the meeting, so—"

"To see Ethan?"

She nodded and bit her lip. We never really spoke about Ethan when we were together. Both of us avoided mentioning his name. Or maybe it was just me.

"Yeah. He's stayed after his term finished to do some work. Plus, it's my birthday on Wednesday."

"Of course!" I said. "Ah, that'll be nice."

"But you're coming to mine and Gabe's party on Friday, right? I'll be back by then."

"Sure." I nodded.

"I could go straight home from Cambridge for the Christmas holidays." She shrugged. "But I thought, I dunno, I want to see everyone that weekend before people go home."

I smiled at her. "I'll be there."

Pushing her lips together, Maya swayed.

We said goodbye and went our separate ways. After a few steps, I looked back at her and watched her figure float into the hush of the opposite darkness.

When I got home that night, I checked my phone and saw I had a missed call from Orla. I got them every now and again and tended to just ignore them.

I ignored this one, too, deleting it and then locking my phone.

Standing still in my room, I felt at a loss knowing Maya was leaving Edinburgh for a few days. I had so much that I wanted to express, and there was only one way I knew how to do it. I sat in my room that evening, lit my candle, and started to write.

17

BIRTHDAY

Maya

Cambridge, England

"HAPPY BIRTHDAY," Ethan whispered in my ear.

He kissed my bare arm and stroked my hair as I woke up and rolled over in bed. His breath smelled of the morning and his eyes were crusty, so he rubbed at them lightly with his fingers. We looked at each other and smiled.

"Thanks."

"Twenty," he tutted and shook his head. "No longer a teen-ager."

"Don't!" I whined, covering my eyes.

"Wait here," he said, swinging himself out of bed. He flung a T-shirt over his head and rushed out the door.

It slammed shut and I lay there in the silence. The curtains next to his bed blocked out most of the light, except for a few speckles which snuck through the cracks at the bottom. I put my hand on the material and opened them abruptly, letting the light burst into the room.

It was a pale December morning. The grand sand-colored buildings of Ethan's college spiraled up towards the sky, which was the color of Aisling's eyes. Towers and metal spikes pushed their sniffling noses up into the air above the roofs. The grass of the college quad was crispy with frost, its tips standing upright like sparkling needles. I could feel the chill radiating from the thin windowpane. I breathed in and out.

Lying back down in Ethan's bed, I picked up my phone from his bedside table. I had a few birthday messages already. I flicked through them, randomly replying to a few. Naomi, Tim, Lucy. They all said they hoped I had a nice day with Ethan. They knew I was there, visiting him.

As I scrolled, I saw Aisling's name. She'd sent me a message at 1:09 a.m. I sat up in bed and opened it immediately.

> Happy birthday Maya—hope you have a really great day. Thanks for being such a great friend. See you Friday. X

I felt my eyes going hot, my nose starting to run. I wiped them harshly. The shell of my skin felt like it was cracking. 1:09 a.m. What was she doing up at that time? Had she been with someone else? I thought about what she'd called me: her friend.

After we'd been to life drawing, I'd gone home and sketched Aisling. I'd closed my eyes and pictured her there. Her long, chestnut hair. Her dark eyebrows relaxed above her bright cobalt eyes. I'd put the drawing under my bed after finishing it. It was weird, what I was doing, I knew that, but I couldn't help it. I couldn't stop thinking about her. Things were starting to spill over.

"Here we go," said Ethan, opening the door with his elbow and bringing in two cups of tea.

Seeing him made me smile, even despite the haziness of my brain. I wasn't sure exactly what I was feeling, but I needed to tuck it away and focus on him. Being with him was good. I cared about him; he cared about me. This was what I wanted, I told myself: I wanted to be with him.

He popped the mugs on the bedside table and clambered back into bed.

"Right, birthday girl, let's talk plans for the day." He leaned over to kiss me.

The insect wriggled inside me, waking up and growing with each second.

Ethan and I walked around Cambridge that afternoon. The bells clanged and echoed through the cobbled alleyways. We ate cake and went out for dinner. I had a lovely birthday with the kind, funny, wonderful boy I'd known almost my whole life. I would figure this out and everything would be OK.

• • •

THE NEXT DAY, Ethan walked me to the station in Cambridge and kissed me softly on the lips. He said he'd see me when we were home next week for Christmas.

"Yeah, only a few days away." I shrugged.

"Have fun at the party tomorrow." He smiled. "Sorry I can't be there."

I told him it was no problem and thanked him again for the amazing day yesterday. We kissed one last time; then I got on the train.

I slept almost the entire journey back to Edinburgh. When I arrived at Waverley station on Thursday evening, the superficial white lights stung my eyes. My bag felt weighty on my body. Weakness in my muscles, all over. Announcements and the bustle of people droned, but I felt a deep sense of relief to be back.

When I got into the flat, I put my bag at the foot of my bed and went straight to sleep. I slept right through the evening and all through the night. Gabe came in at one point to see if I needed dinner.

"I'm OK, thanks," I murmured, rolling over.

I woke up the next morning, Friday, at ten a.m. I'd been in bed for fifteen hours. I stretched my toes down over the edge of the bed and then curled up in a ball. Rubbing my eyes, I lay there and my mind drifted. I got the drawing of Aisling out from under my bed. I looked at it. After a few moments, I shook my head and tucked it away again.

The rest of the morning, I sat at my desk, peering out over the street. At some point, amid my daydreaming, I began to write. It started slowly, just a few words, but as I sat there, it started to erupt from inside me.

At lunchtime, Gabe came in and asked me if I wanted something to eat.

"Maya, I mean this in the nicest way possible, but you look awful."

"Thanks," I said, typing madly.

"It smells a bit funny in here."

"I'll shower before the party."

"OK, good," he sighed. "I'm glad you haven't forgotten about that."

I heard him rolling his eyes.

"You know it's tonight, right?"

I nodded.

"Lunch?" he shouted at me as he left the room.

"I'm alright. Thanks, though," I shrieked after him.

Gabe brought me some pasta anyway. I wasn't hungry, even though I hadn't eaten since lunchtime the previous day. By the late afternoon, I'd only made trips to the loo and to get water. I stank and my bag from Cambridge still wasn't unpacked.

Eventually, with an hour or so until the party, I stopped typing. I knew then I'd never show it to anyone, and I knew I didn't even really understand what I'd written, but it was there and that was enough.

I got up from my desk chair and flung open the windows in my room, letting the cold, dark air sink in. Then I got my towel off the hook and went to have a cold shower. I let it wake me. As I got out, my teeth chattered, but I felt refreshed. I spent time getting ready. I put on my makeup and slipped into my sparkly silver dress which draped over my boobs and down my back.

"Wow. Maya, alive at last!" Gabe twittered as I came into the sitting room. He was putting up fairy lights and balloons. I twirled around.

We drank gin and tonics as we put up the decorations to music which blasted from the speakers until the floor shook. Gabe gave me a personal rendition of "Upside Down" by Diana Ross while standing on our coffee table and dancing along. I sat and sipped my drink, cheering him on.

Around eight p.m., people started to arrive. They kissed me on the cheeks and hugged me, passing me bottles of booze, which I added to the communal table. I drank. The mesh of their perfumes and colognes wafted into the increasingly humid room. I drank more. The sound of music and laughter bubbled through the flat, building and building. I drank more still.

Time flew by, but Aisling still hadn't arrived. I downed a glass of wine and went to the bathroom to sit on the loo and check my phone. I viewed the last message I'd sent her on the train back from Cambridge.

> Just on the way back to Edinburgh! Hope
> the last meeting of term goes well tonight,

sad not to be there. You're coming
tomorrow, yeah? X

No reply. I read it over a few times as I sat there. I typed out another message to her. My vision was slightly blurry, so I had to close one eye to stop the letters from moving around the screen. Once I'd checked for spelling errors, I clicked send. Then I got up from the loo, flushed (even though I hadn't peed), washed my hands, and assessed my face in the mirror. I was more drunk than I'd realized. My eyes looked drowsy, and my makeup was smudged, so I rubbed the smears from under my lashes. As I came out the bathroom door, I observed the people. Dancing, drinking, smoking out the windows.

Then I saw her.

She was over in the corner, talking to Harry and Isla, nodding her head as they rambled on. She had her hands in the pockets of her long black coat. My heart started beating fast.

I turned away and walked straight to the kitchen, making a beeline for the sink. I turned on the cold tap and filled up a glass, gulping the water down. I couldn't let her see me like this. I was so drunk. I wasn't even sure what I looked like anymore. I drank two more glasses of water, then plonked the cup down on the counter and gasped for air.

"You alright there?"

I swiveled around. She was looking straight at me, like a vision.

"I got your text just now." She narrowed her eyes.

Then I felt it. Coming up. My stomach connecting to my mouth and everything in between. I couldn't stop it. I ran to the bathroom, covering my lips and retching. After slamming the door shut, I vomited straight into the toilet. My body convulsed and shuddered as it pumped out the poison. I wasn't completely present, but I felt Aisling come into the bathroom behind me. She shut the door and put her cool, dry palm onto my sweaty forehead. My whole body was clammy. I threw up again. Tears started to stream from my eyes; my nose started to dribble.

"I'm so sorry," I cried. My throat was hoarse from the vomiting. Even more came out.

"You're alright." Her soothing voice got under my skin.

Aisling gathered all my hair up and gently pulled it back behind my ears, then put her hand on my forehead. Her touch was so cool and collected, and even in that moment it made me tingle.

After it had all come out, I curled up on the bathroom floor while Aisling went to get me some more water. When she returned, I was sat against the wall with my head tipped back, trying to cover as much of my body as I could with the dress. I was shivering. Aisling sat on the side of the bath and fed me some water from the glass. My throat felt raw and my lips were dry and cracked.

"I think I want to go to bed."

"Come, I'll take you."

Putting down the glass and standing up, Aisling offered me her hand before helping me up. I stood and brushed my teeth while she waited. The mirror above the sink captured her reflection. We studied each other's faces, and as we did, I saw a look of hers that I'd never encountered before.

While she went to get me more water, I walked to my room, took off my makeup, and got changed into my pajamas, then snuggled myself under the duvet. My eyes felt so achy I couldn't open them. My head throbbed, and my stomach felt empty, but relieved.

I heard Aisling come in and put the glass on my bedside table. The way she breathed was sweet and melodious. Even with my eyes closed, I could imagine what she was doing. She was sitting there, watching me. Skin glowing, eyes dancing like flames. It made me feel safe and cocooned knowing she was there.

I felt tears start to roll down my cheeks. I couldn't bury this anymore or hold it in any longer. I had been denying things to myself for so long.

18

THE DISTANCE BETWEEN THEM

Aisling

Edinburgh

I SAT ON a frost-covered bench looking out over the meadows. It was the morning after Maya's birthday party. An icy layer of mist hung delicately above the sparkle-tipped grass. The trees clenched with the cold, their branches like thin arms, contracting and bending into their own bodies.

I was alone, my only companions the trees and the surrounding hills. The only sounds were from the birds.

My entire body was covered in layers and layers of clothing, except my face, which stung from the cold. I wore my thick black woolen coat, black gloves, and a green fluffy hat, which was a hand-me-down from Mary. I was holding two takeaway coffee cups. One for me, one for Maya when she arrived. As I sat there and looked at the view, I enjoyed the sensation of the warm liquid seeping through the cup, through my gloves, and onto my chilly hands.

I was flying back to Clare later that day for the Christmas break. Back to the grayish streets and boggy fields. It made me feel sick, the thought of going back and seeing my parents. Seeing

Ma. We hadn't spoken in months, and I was glad of it. She'd send me the occasional text. Cold and simple, like an instruction to send a relative a birthday note. Nothing more.

In Edinburgh, I knew I'd been blocking everything out, becoming distracted with Maya and burying my past deep into the earth. I had been free here; life had been good, and so I felt a strong sense of dread about returning home.

When Maya had gone to Cambridge to see Ethan, I'd stayed inside for four days straight, just reading and trying to get ahead with my work for the following term. I hadn't showered, exercised, or even eaten, really. Karen came to knock on my door and ask if I wanted anything from the shop. I could see her grimace at how messy and stuffy my room was.

"I'm alright, cheers," I'd said, knees curled up on my desk chair.

"You look tired," Karen whispered, her curly, straw-like hair bobbing on top of her head. "Are you sure you're OK?"

I'd nodded at her with my lips closed. Karen had started to shut the door.

"Thanks, though, Karen."

She left the door ajar for a second before closing it completely.

The night before Maya's birthday, I worked for hours, and eventually it passed midnight. I texted her in the small hours of the morning to say happy birthday, right before letting myself move over to the bed and drop off for a while. After some broken sleep, I woke up at five thirty a.m. Wide awake. I got up, pulled on some clothes, and strolled to the meadows.

In silence, looking out at the emptiness of the field, I reflected on what it felt like to be around her. When we were together, I became exactly the person I'd always wanted to be. I could reject the words Ma had whispered in my ear. I could somehow shove away my history; the guilt, fear, helplessness that my mother had always made me feel.

As I sat there by the meadows, I finished writing the poem, the one I'd started after we went to life drawing. It came out of me like water flowing in a fast-rushing river. Over rocks, swirling in whirlpools. Then I walked home just as the sky splintered with bright silver light coming through the clouds. I made coffee and sat at my desk. The wafts of steam glided up in silver ribbons. I threaded the poem together for one last time. It felt like hanging up a picture on the wall or stringing up a row of fairy lights. It wasn't perfect, but it shed light; it allowed me to see new shapes, new colors.

On Friday before the party, I dressed all in black (as usual) and didn't bother putting on any makeup. I'd accepted the fact that I didn't have a natural talent for it. My phone had pinged just as I arrived and was saying hello to Harry and Isla. It was from Maya.

> Ash, where are you? I miss you. I really
> want to see you.

My stomach felt as if a thousand horses were galloping around it.

Then I saw her.

That night, watching her get sick, the smell of the alcohol, her slurring words, her eyes in the mirror, it had reminded me of my mother. I'd tucked that thought away immediately and pushed it down to the pit of my stomach. It wasn't my mother; it was Maya.

I put Maya to bed and watched her fall asleep. I saw the tears dribbling down her cheeks and onto her pillow. I thought about what it would be like to get into that bed with her, to have her body close to mine, to run my hands through her hair or over her back and arms. What would it feel like to kiss her cheeks, her neck, her mouth?

Briefly, I took my eyes off her and cast them around the room. Coloring pencils on the carpet. A chipped vase with dead flowers drooping and wilting like sad faces. Clothes strewn over the furniture. I thought about what they'd look like on her body. The colors next to the complexion of her skin. There were mugs on the desk and a dirty bowl still full of pasta. No wonder she'd gotten so drunk, if she hadn't eaten.

Then I caught a glimpse under her bed. I saw drawings in black and white, a bit like the ones we'd done at the life drawing class together. These were different, though. Maya, I could see, was fast asleep, so I leant over and carefully picked up the drawing at the top of the pile. I stared at it. A girl, lying naked. Her hair was long, hanging over her breasts and torso. The person. The girl. My brain blipped and then snapped.

Was this me? Was this a drawing of me, done by Maya?

Maya was not only a good writer, but she was also exceptionally good at drawing, and this face looked exactly like my face. The more I regarded it, the more I thought it must be me. My brain turned to sludge and my heart pumped so hard I thought it might explode. I tried to figure out what this meant. I hoped I knew what it meant, but the last time I'd hoped, I'd been disappointed.

I looked at Maya, sleeping. I put the drawing back under her bed, checked she was OK, and moved her into the recovery position for good measure. Afterwards, I went to tell Gabe what was going on and asked him to check on her in a bit; then I left the party.

I sat there on the bench by the meadows the morning after all this, waiting for her.

She'd messaged me in the early hours after she'd woken up from the sound of people leaving. I'd been lying awake, not sleeping, so I'd messaged her straight back and we'd arranged a time and place to meet. I was leaving to catch my flight that afternoon, so early this morning was the only time we could see each other. She'd said that was fine; she wanted to see me.

I turned and saw Maya coming from a distance. She cut through the fog with long strides. Her aura, powerful, sending shock waves through the air. As she approached, she took one of her hands out of her coat pocket and waved at me, lifting it limply, giving a faint smile. When she got close to the bench, she stopped and held my gaze, then sat down. I could see her eyes were puffy and red, perhaps from crying. She wasn't wearing any makeup, and she smelled of alcohol, dry shampoo, and perfume. Beneath her coat, she was wearing an oversized hoodie, leggings, and trainers. There was something intimate about seeing her this way; it made me feel that she trusted me. She wasn't trying to hide things from me anymore.

"Coffee."

"Thanks," she replied, taking the cup.

I left space for her to speak.

"I'm so sorry." Her voice was husky and raw.

"What are you sorry for?" I laughed.

"Everything. I'm sorry about last night, sorry about my message, sorry about all that stuff a few weeks ago, sorry about the state of me right now."

"You don't need to apologize, Maya. You've got nothing to be sorry about."

As she twisted to look at me, I turned to her. Our faces felt close. I was aware of the distance between them.

I asked her if she felt OK this morning. She said not really, but that she'd be fine later. She apologized again for being so drunk last night and thanked me for looking after her. I just smiled, softly, and told her it was nothing. She didn't know that I would have done anything for her.

"Aisling."

Her saying my name was like music. It held an entire melody, the quiet parts, the crescendos, the clashes, the harmonies.

"I've been feeling very confused recently."

I nodded, and hope glistened in my stomach. "About?"

Her gaze left mine and retreated into her lap. "I think you know what I'm confused about, Aisling."

We waited and watched the frost undress itself from the grass.

I knew I had to give this one last try. I had to lay my cards on the table.

I took the folded-up poem that I'd written out of my pocket. My heart raced as I passed it to her, but I had to do it. Placing it in her hand, I held my fingers there for a moment. I let my thumb stroke the side of her palm briefly; then I withdrew it.

Maya's eyes looked glossy, a deep chestnut brown. Turning to the paper, she put her coffee down beside her on the bench, and opened it, unbending the folds. Her eyes moved over the words. Her breathing got even heavier. Chest rising and falling. I watched her. Eventually, she looked straight ahead, and folded the piece of paper back in half.

"This is beautiful."

Time crawled by like a caterpillar biding its time.

"Is it about me?" Her voice wobbled, as if it were a string being plucked.

I paused. "You know it is, Maya."

She buried her head in her hands, and I could hear her starting to cry. I put my coffee down, too, then moved my hand onto her back and, slowly, wrapped my body around hers. I held her close, and she held me back. We fitted together, and I shut my eyes as I felt her trembling. Gradually, she steadied like a ship through a storm, and we released one another, remaining close but with space, unnecessary space, emerging between us.

"Can you give me some time? I need some space to . . ."

I nodded, relieving her from explanation, assuring her that I understood. I understood her, completely. She asked me if we could speak again after Christmas, once we were back. I said I would wait for her to contact me when she was ready.

We sat there in silence.

"It's stunning here, isn't it?"

"It really is," she whispered.

19

THE MISSING CONNECTION

Aisling

County Clare

THE FOLLOWING DAY, I woke up in Clare. Everything felt so familiar—the sounds of the house, the dust on the toys, the religious paintings, the bits of clothing sticking out the wardrobe—but I felt no connection to it. I was more different than ever, stronger than ever. Even my body felt more muscular from all the running.

I looked at the clock on my bedside table. It was 8:45 a.m. and it was a Sunday. Half an hour until we left for Mass.

In the shower, I scrubbed my face with the same cracked tea tree soap as before, surrounded by the same flesh-colored tiles and the pale pink bath. Afterwards, I put on a new dress, one I'd bought in Edinburgh. It was black (of course) and long.

I looked in the mirror and let my hair down from the bun which I'd put up to shower. It was fluffy where the droplets had caught it. I flicked and smoothed it down with my hands so it ran over my collarbones. Inspecting my face close to the mirror, as I'd done so many times before, I realized I looked less afraid, and

I was. I put on a pair of hoop earrings and took one last look before leaving my room to go downstairs.

As I descended, I heard the familiar sound of the classical music. I gently swung open the door to the kitchen. The same as before I left: newspaper, loaf of bread, coffee, and orange juice perched on the table.

My three siblings, Sean, Jack, and Mary, weren't coming home from Dublin until Christmas Eve, because of work. Pa had told me this when I arrived back, saying it was just the three of us for now. This was confusing to me. Ma had texted me saying that everyone would be back the day I arrived home. That's why I'd booked my flights for the Saturday. It made me fearful, like she'd done it on purpose. She was a wolf chasing one lone deer away from the pack.

"Morning," I said.

As ever, my mother didn't turn to greet me. She was mopping the floor. It wasn't the time to do it, but she was doing it anyway. Sloshing the water around as she dunked the jellyfish end into the lemony liquid, twisting and turning it, wringing out its tentacles, then frantically wiping the floor even though it was already clean.

"Glad to see you're ready for Mass on time. How does it feel to be back?" asked my father without looking from the newspaper. One of his legs was crossed over the other, protruding from the realms of the table.

I sat down and started to slice myself a bit of bread from the loaf. Sawing into the thick, starchy block.

"Strange."

"That doesn't sound very positive," Mother said.

She propped the mop up against the wall and plucked at the fingers of her washing-up gloves. I stood and popped my two slices of bread into the toaster, then pressed down the button.

"I don't remember buying that dress for you or Mary."

"I got it in Edinburgh."

Ma pulled a snooty, judgmental expression, her bottom lip jutting out and her eyes widening. Pa started to fold up the newspaper.

"Time to go."

I stood still in the kitchen as they moved to leave. Picking up what they needed, shuffling things around. Ma hung up her apron and put her washing-up gloves away; Pa placed his empty, pulp-lined orange juice cup in the sink. I looked at the table. The orange juice carton, the folded newspaper, the bread.

My toast popped up.

"I don't think I'll come to Mass this morning."

They were out in the corridor. I couldn't see them, but I heard their footsteps stop. Their breathing pause. I heard them exchange a look.

"I'll wait in the car," my father whispered to my mother. He retreated to the door, and it closed behind him.

"You'll be coming to Mass, Aisling," Ma instructed from the corridor. Her voice was deep and contemptuous. A slide on gravel.

"And what if I don't?"

I stood still in the kitchen. Still. Still unable to see her. My heart started to beat. Faster and faster.

Very slowly, I heard her stalking towards the kitchen, towards me. Her ominous figure appeared slowly around the door frame, shifting into focus.

"Aisling," she said very slowly, "get in the car."

Her head was dipped so that her pupils filled her eyes like blood. I shook my head.

"No."

There was a moment where our breathing hastened. It was like a mirror.

Ma strode towards me like a tiger hunting its prey, bearing its clenched teeth, her hair straggling out of her pristine, combed-back bun.

I'd known it was coming. I always did.

She drew back and lunged at me.

The back of her hand went to hit my face, but as it came closer, the stone of her ring snagged on my earring. It yanked it, ripping the earlobe in half, snipping it like with a pair of scissors. It was as clean as a curtain being torn in two.

I screamed, yelled, winced as I dropped to the floor.

Feeling the warm blood flowing from the cut in my ear, I could sense it trickling down my hands, then my arms.

Ma stepped back and covered her mouth. She was shocked, but not because I was hurt. It was the fact it hadn't gone to plan; that's what had shaken her.

I brought my hand down. Drops of blood distributed like leopard print on my pale palms. I grabbed some kitchen towel, unrolled as much as I could, and cupped it around my ear. I squeezed. It stung. My breathing was heavy and uneven. I tried to inhale and exhale slowly, but the tears came to my eyes. I blinked hard, retracting them as much as I could, but they stung like wasps.

As her shock subsided, my mother's stern facial expression resumed. She grabbed my arm and yanked me through the house, out the door, then shoved me into the car.

My father drove us through those familiar streets. He was pristine and uninvolved. He didn't even look back at me. He'd known this sort of thing to happen my whole life, and he had never done anything about it.

I couldn't observe the swampy fields or the dead hedgerows that whizzed past the car. Kitchen towel still in my hand, I just moved it around to get the bits which weren't covered with blood. I bit my tongue. The pain hummed in my ear. I could feel as I pressed against it, the slit. The gauge of flesh. The missing connection.

I need to go to a hospital, I thought, not a fecking church.

Why did they want other people to see me like this? To see me bleeding? Perhaps they relied on no one ever saying anything. Perhaps they were insane. Perhaps they didn't care and felt that missing church was worse than injuring their daughter.

As we pulled into the car park, Ma rummaged in her handbag and pulled out a thin plaster.

"Here."

I looked at the plaster. I shook my head, the sweat glistening on my brow.

"That won't work." I almost laughed. "I'll just stay in the car."

My voice shook from the pain. She turned around to me as Pa reverse parked into the last remaining bay.

"Just put it on."

She handed it to me, and I did. As we walked into church, I cupped my plaster-covered ear with a tissue. Everyone was singing the hymn. Ma and Pa joined in as we shuffled into the back row of the pews.

• • •

LATER THAT MORNING, I sat on my bed, feeling sick from the pain of my earlobe. I was desperate for someone to help me. I would have just taken myself and caught the bus to the hospital, but I didn't think I could manage it.

I'd texted Orla. I was sure she'd be back for Christmas by now, and she was the only person I could think of who would understand but not ask questions.

I'd told her that I had cut my ear and asked if she could help. She'd responded almost immediately, telling me that she'd meet me at our spot in ten minutes. I felt such relief.

The spot was where we had met while we'd been seeing each other. A few roads down from my house, at the end where no one would be able to spy. I walked slowly to meet her there. I didn't care where my parents thought I was. Seeing the silhouette of Orla's thin figure in the car made me want to weep.

I got in and shut the door.

"Cheers."

"Don't mention it."

She pushed my hair behind my ear and analyzed the cut, flinching as she saw it.

I told her please not to ask me what had happened, and she nodded reluctantly. She told me we'd just get it sorted first. I felt calm for the first time in days.

Orla took me to the hospital, and we waited for a little bit. Eventually, I got local anesthetics before my ear was stitched. In the mirror, when I looked at it afterwards, it was yellow with big red ladybird dots. I felt dozy and weak, unable to distinguish reality from dream.

When we left, Orla drove us to Spanish Point. I hadn't wanted to go home yet, so she suggested we go sit somewhere for a while, handing me a sandwich to eat, telling me I must be hungry. She'd made it for me before she picked me up. It was the kind of care that I had never had from my family, only ever from her. Just like the time she'd bought me that bacon roll to school.

We drove in silence. I ate the sandwich. She was right, I was starving, and I watched out the window as we glided along the shore. The sky hushed the pale green grass, and the gray water cooled it, lapping at its fuzzy edges. I thought about Seamus Heaney's poem about Clare, "Postscript." I thought of how he speaks of the relation between the wind and the light, of how the wild ocean glitters and foams. The poem came to life as I observed the place. I did feel caught off guard by this whole thing, just as Heaney had said.

We sat at a deserted part of Spanish Point in the long grass for a while, looking out to the sea, sipping at hot chocolates that we'd picked up on the way. Orla had told me it would help the shock and the sleepiness. The air was freezing, but the sky was settling into a peachy haze. Pastel light brushed over the base of it like a watercolor, and lemony light streamed through the cracks in the clouds. The wind blasted us, skinning our cheeks.

"How're you feeling?"

I looked out to the sea. "Better. A bit. Thanks for doing all this. It's good of you."

I swallowed, trying not to cry. I could feel Orla studying me.

"I'm sorry I ran away, Ash, last summer."

I turned to face her. She kept talking, but broke eye contact with me, refocusing on the horizon.

"You've probably noticed from all my drunken calls that I regret it. I miss you. There's—there's no one like you."

I didn't know what to say, so I didn't say anything.

We sat there in silence for a while longer, just looking out at the lapping waves, the changing tides.

"I think I'm still in love with you," she whispered.

I turned my face towards her. She looked at me then, her green eyes darting around. Orla leaned in and kissed me, gently, on the corner of my mouth, just as she had that time outside the changing rooms. I let her.

As she drew back, Maya flashed into my head.

What was I doing?

But no, Maya had asked for time. She wanted time to think,

and, really, she was still with Ethan. Perhaps, after all my waiting, I'd only be disappointed again.

I kissed Orla in return. Her scent and taste were so recognizable and consoling. I longed to feel loved in that moment, and Orla made me feel that. I put my hand behind her neck and drew her in, and in, and into me.

• • •

MY PARENTS DIDN'T comment on the fact my ear had been stitched. It was like nothing had ever happened. The same pattern as ever, over, and over, and over again.

On Tuesday, Christmas Eve, Sean, Jack, and Mary arrived home from Dublin together. They'd all driven in the car which Sean, my eldest sibling, had just bought with the money he'd been saving from his job. A dark blue VW Golf. As they paraded in through the door, we greeted one another in turn.

The stitches had left a big scar on my ear, very much visible when my hair was tucked back. Jack and Mary both looked at me as they entered. I'm sure that they saw it, but they ignored it. They must have known it was from something they didn't want to hear about, else they would have asked.

Willful ignorance was the norm in this house. I had always known that was the case for Ma's drinking, and apparently, even now when we were adults, it was the case for any harm that she caused me as well. When I'd been young, the marks were covered by clothing. But my ear, it was obvious, and it was still ignored, just as my hand had been last Christmas. I was sure Ma had spun a line to them before they arrived, providing another flimsy explanation.

Sean had taken me to one side, though, as soon as he got through the door.

"You alright?" he asked me quietly in the corridor, without making much eye contact. The others were all in the kitchen by then.

"Yeah, good. How come?"

Sean had flared his nostrils and looked directly at me, pointing towards my ear. My instinct was to touch it, but I held back. I frowned at him.

"Did they have something to do with that? I didn't say anything to them about what I saw, you know, in the summer, I promise you I never—"

I cut him off. "I know, I know you didn't, Sean."

I felt he was realizing, slowly, what had been clear for so long. But I knew he was still scared of her, just like the others. I gave him a look of finality and he dropped it soon after that. There was no point in dragging him into it.

For the following week or so, the house felt crowded, like it always did when the whole group were home. It changed when the others were back, from a nightmare into some performance of a family. I felt safer when everyone was around, more watched.

Sean and Jack shared their old room, and Mary slept in her room just next door to mine. The house smelled of cinnamon, Christmas tree needles, and candle wax. It was filled with classical music and loud voices talking over one another from when the sun came up until late into the evenings.

The night my siblings returned home we had a big meal around the dining table. Then, in the evening, we all piled into the big Citroën and went to Mass.

The church looked so festive, as it always did this time of the year. Decorated with lights and ornaments. Church felt normal on days like this. We all sang the hymns, mostly out of tune, and read together the prayers and the liturgy. One family. In a line.

Only connected by these words we spoke in time, together. We all listened to the readings and the homily, with varying degrees of concentration.

Afterwards, we piled back into the car, wrapped in coats and gloves and scarfs, rubbing our hands together from the cold, switching up the heating. I was quiet in these situations, but at least I didn't feel alone. When we got back, we all went into the sitting room to light the candle. The room was dusty and stagnant as usual, and the small Christmas tree lights flashed in the corner, throwing glints of light onto the multicolored baubles.

"Alright." Ma clapped, pushing her way through, holding the big white candle.

Sean was yawning, Jack was chatting rubbish to no one, Mary was rearranging the decorations on the tree, Pa was getting something to eat in the kitchen. And I stood in the corner, silent.

"Mary's got to light it, else it's bad luck." Ma leant over the sofa and placed it on the windowsill. "Cormac," she screamed, "Cormac, get in here, we're lighting the candle."

Pa appeared round the edge of the door, stuffing a cracker with cheese into his mouth. Mary ruffled her feathers and picked up the matches from the mantelpiece, strutting over to the candle. She lit the wick, and the flame flittered like a small blue-and-yellow tadpole, dancing and swiveling its tail.

"Ah, lovely. Can I get to bed now?" requested Sean.

"Alarms for Mass tomorrow," announced Mother, raising her finger in the air, leaving the room.

I sat and watched the flame while everyone else went upstairs to bed. Gradually, the house became silent.

As I sat there, the clock ticking in the background, I thought about Maya.

I'd seen Orla once more since she took me to get my ear

stitched. The day after, on the Monday, she'd messaged to ask if I would want to see her again. I was lying on my bed at the time.

I thought about Maya at that point as well. I wanted to see her. I was missing her, thinking about her every day. We hadn't spoken since I'd seen her at the meadows in Edinburgh and handed her that poem. But that was what she'd wanted. She'd wanted space. I thought about her, back at home at that moment, and the image of her and Ethan together flashed into my head, making my stomach turn.

I messaged Orla back telling her I'd be free to see her that afternoon.

Later the same day, I told Mother I was going for a long run, and that I'd be back in an hour or so.

"A run?" Ma had spat. "Why on God's earth would you do that?"

I'd left and met Orla at the spot, and she drove us to a secluded location just as she had done when we were together. The sea whipped in the wind, salty sprays of water flicking onto the car windscreen. We didn't talk much. When we got there and parked up, she turned to me, and we started kissing. Her lips tasted as they always had; her skin was soft and cold. She moved her body over towards me, on top of me, and laid the seat back. As we had sex, I felt hollow. I thought of Maya the whole time.

I felt a single tear drop from her eye onto my cheek as we lay there afterwards, our bare bodies touching, shivering. Connected but detached.

I didn't want this. I didn't want to be with Orla. This had been a mistake and it hadn't been fair to her. I stayed quiet as Orla drove me back to the spot. I apologized to her when we got there. She must have sensed how I felt because she bit her lip and told me not to be sorry. Her words flew over my head like an eagle. I felt terrible. When I got into the house, I took off my running trainers and went straight upstairs. I locked myself in the bathroom and stared in the mirror, trying to ignore the fact that my frowning, forlorn face made me look just like my mother.

I observed my lips, which Orla had kissed only an hour or so ago; I studied my ear, still yellow and freckled with red dots. I would have cried, but my body felt numb, although my heart was cracking.

I didn't get in touch with Orla again that holiday.

I must have watched the candle flame for an hour or so as I lay in the sitting room on Christmas Eve. I thought and thought until my eyes started to droop. Then I went up to bed, nestled myself into the sheets, let out a big yawn, and fell asleep. In the morning, I woke to the sound of the shower running and the pipes screeching and the usual, repetitive routine of Christmas Day.

20

HOME

Maya

London

ON THE TRAIN from Edinburgh to London, I gazed out the window the entire journey. The carriages snaked around the shore of the lashing sea. Moss green beneath steel blue, water-filled clouds. At each station, I observed loved ones greet one another or part ways as they stood next to golden-glowing Christmas trees or beside café tables. They exchanged passionate kisses, short pecks, long hugs, rushed embraces prematurely broken. I watched the way people's faces changed when they parted from others. A big sigh, a fond smile, or perhaps even a concealed sense of relief.

What had I looked like when I said goodbye to Aisling on that bench a few days before? Had anyone seen me then, in the same way I was seeing these strangers?

After Aisling had left me there, I wept as I sat on that bench. I watched her walk away, the swaying motion of her body, her hair draping down her back. I felt broken in two. I wanted to run to her, look into her eyes, but I knew I needed time to process everything and to speak with Ethan.

The day after my birthday party, I reread the poem that I'd written after returning from Cambridge. Reading it, I had realized that it was so blatantly obvious how I felt. It was so stark in my words. I had been denying my feelings for Aisling since I'd met her. I had fallen for her, and I needed to face that rather than trying to hide it away.

I watched families from the train window. Little children with bows or clips in their hair, kids with big woolen Christmas jumpers. They held their parents' hands. One man, one woman.

For most of my life, this had, overwhelmingly, been the model of a relationship that had been normalized to me, along with certain ideas of beauty, success, friendship, and family. But why? What if I didn't subscribe to these models?

When I got to King's Cross station, the jingle of the holidays was everywhere. Music flowed. The big Christmas tree stood tall, covered with flickering lights. I caught the tube home with my suitcase, lugging it down the stairs and rolling its squeaky wheels through the tunnels. As I sat there, I watched tourists analyze maps, tap their lips, furrow their brows. I watched people on their phones, bopping to music or scrolling mindlessly.

From Shepherd's Bush Market, I pulled my suitcase through the streets. Past the dry cleaners, past the stalls, past the smells of hot cocoa.

When the door to my parents' house swung open, I entered and breathed in. I closed the door behind me and stood looking into the silence.

I sauntered slowly around the house. The sitting room, with firewood stacked up, the countless pots and glass ornaments. The corridors, where the walls were covered with Mum's paintings. The creaky wooden panels of the floor. Each of the noises and smells were so familiar to me. In the kitchen, copper pots and mugs hung on hooks. The smell of rose petals, star anise, nutmeg, lemon. The washing machine whirred. I went towards it and sat on the floor in front of it. The clothes went around and around in the drum, the bubbles of laundry liquid crackling. I had loved doing this as a kid, just watching it spin, smelling the freshness, knowing that something dirty would soon be clean.

"Maya."

I turned. Mum stood in the doorway to the kitchen, her glasses pushed up onto her head, her eyes tired, her overalls dirty with paint.

"Mum."

"What are you doing, my sweet?" she laughed. "How long have you been here, I thought you were due back at three o'clock?"

I scrambled up off the floor.

Before I could think, I ran to her and buried my head in her chest. As I hugged her and held her tight, I started to cry. She didn't say anything; she just held me. I could smell her. Paint and fresh linen. Eventually, I caught the sobs in my throat, and they slowed. I relaxed, swallowed, drew back.

"Darling," she murmured, holding my face in her rough hands, looking into my eyes, "what's wrong?"

I gazed around the room, trying to search for words, but she focused on me, untangling my expression.

"Do you want to talk about something?"

I nodded, biting my lip to hold back more tears. "If you have time."

"I think the painting can wait." She smiled.

"Is Dad home?"

She shook her head. "Work."

I sat at the kitchen table in silence as she boiled the kettle and put some fresh mint into two mugs. She poured the water over the leaves and a crisp smell diffused through the air.

Once she was sat and had handed me the mug, I spoke slowly. I told her everything from the beginning. I described Aisling to her, and I explained to her, as honestly and clearly as I could, how I felt. Mum nodded as I talked, asking the occasional question, but otherwise listening intently. It felt like a relief to finally articulate my feelings. I hadn't spoken to anyone about them. Gabe had tried to talk to me the day after my party, insisting that he knew something was up, but I'd politely brushed him off.

Once I'd finished, I took a big breath. Mum wiped her eye with her forefinger, resting her head on her other hand. I waited for her response as my stomach simmered.

"It sounds like you might be falling in love with Aisling."

I nodded, biting my lip.

Across the table, she outstretched her arm. Her palm and fingers molded around my hand and squeezed it tight. I watched her face smile. Her eyes sparkled and her forehead creased as she understood what I was trying to say to her. I could tell her response wasn't coming easily, but it was brewing.

"I am so proud of you for talking to me about this," she said after a moment.

I squeezed her hand back.

"I love you whoever you are, whoever you love," she continued, moving her other hand onto mine.

"Mum," I muttered, my throat weakening.

We stared at each other, and with my eyes, my expression, I said thank you.

"What about Dad?"

She moved her hands away and used them to rest her head,

her elbows propping her up on the tabletop. Her eyes moved around the room; her mouth pressed into a small round wrinkle.

"Perhaps let me talk to him before you do. I think he will—he will come around."

Reluctantly, I nodded.

• • •

THAT EVENING, I stood outside the big black door to Ethan's family home in Belsize Park. I thought about how many times I'd entered and exited this door and how different I had been then. Primped, preened, fake, proud, arrogant.

I knocked.

Almost immediately, Ethan opened the door and poked his face around the frame, his dimples pressed firmly into his cheeks. It looked like he'd just got out the shower. His face was blotchy and red and his hair damp at the tips. A fresh blue T-shirt, which smelled of washing powder, hung off his shoulders, and gray joggers hugged his muscular legs.

"Oh, Maya! Thought it was our takeaway delivery." He grinned.

"Sorry to disappoint," I said, nervously.

He laughed. "Don't be silly, I just didn't know I was seeing you tonight? Get in here, it's freezing."

I entered the hallway with the fancy floor tiles and the cream radiator covers.

"I should have called or messaged, sorry, I wasn't thinking, I—"

"Maya, it's fine, it's great to see you, you know you're welcome whenever." He shut the door behind me.

"Is now a bad time?"

"Not at all, it's great." He smiled, moving close to me, tucking some loose hair behind my ear. "My parents and I have just ordered a takeaway, but there should be loads if you want some." He gently rested the palms of his warm hands on my face and kissed me. I took a little longer than normal to open my eyes once he'd withdrawn.

"What's wrong?"

"Do you have time for a chat?" I asked him.

Ethan dropped his hands from my face and frowned at me, but he nodded. He led me upstairs to his room. I sat at his desk chair, and he closed the door before sitting on the edge of his bed and resting his elbows on his thighs.

I studied his face in the silence.

The boy who I'd known since I could remember, who I'd watched grow up. The boy who I'd made mud pies with, slid down slides covered with washing-up liquid with, who I'd watched pull pranks on teachers. The boy who had asked me to marry him when we were six. The boy who I had watched sprint like a gazelle through the playground, or run around the track, or play in a rugby match, getting coated with rain and mud. The boy who was kind and gentle, who had always cared about me, who had stayed in my life no matter which school we went to or which city we lived in. The boy who had always been there for me. The boy who I'd seen drunk in clubs with his eyes rolling backwards into his head. Who I'd watched kiss other girls and felt so jealous that my stomach throbbed. The boy who had made me laugh so much my stomach ached. The boy I'd been on holiday with, sung songs with, loudly and out of tune; who I'd leaped into the sea with, or jumped into the pool fully clothed with.

"What's up? Is everything OK?"

I waited for the words to come, and eventually, they did. I told him that I would cut to the chase. I explained that I was struggling to come to terms with my sexuality, and that I wasn't in a place right now where we could be together. I apologized repeatedly. I said I hoped he knew how much I cared for him.

Ethan rubbed his hands over his face. He stared at the floor, eyes wide, breathing slowly but heavily.

I heard the doorbell. It was the takeaway arriving.

"I'm sorry, do you need to go and—"

"It's fine, I'm not hungry. I'll eat later."

His parents shouted his name up the stairs, telling him the food was here.

"I'll be down in a while," he yelled back, his voice catching in this throat.

I felt like a horrible person. The insect was more active than ever, eating my organs and blood and muscles, and it had grown, becoming large and overpowering. I was becoming it. I was the insect. It was me.

I shouldn't have done this now; maybe I shouldn't have done it at all.

Ethan cleared his throat again and sat upright. Then he made eye contact with me. I'd known Ethan for fifteen years, and yet I'd never seen him cry. He wiped his eyes with the back of his hand and sniffed.

"I'm sorry," I whispered to him. "I'm so sorry."

"I knew."

"What?" I asked him gently. "What do you mean?"

"I mean, I had a sense that this, or something like this, might be going on. You'd been so distant with me this term. I knew something was up, but I didn't want to admit it. You're my best mate and I really do love you. I don't want this to end but—" He paused. "I do want you to be happy, and I know this can't be easy for you."

I stared at him. The insect shrunk slightly. It flitted. But mostly it was replaced with a swelling feeling, like water bubbling up inside me. I just breathed as we held each other's gaze. I loved him, too, but not in the way I now knew I could feel. I started to sob. I didn't know how I had any more tears left inside me, but they overflowed.

We both stood and moved towards each other. We embraced. Ethan was my friend in that moment, and it was exactly what I needed him to be.

"Thank you," I whispered to him as I made his T-shirt damp with tears.

I felt little shock waves going through his body as he, too, let the emotion wash through him.

"It's alright," he whispered back. "Thank you for being honest with me."

• • •

WHEN I GOT back home, I clicked the door shut behind me and stood there. I closed my aching eyes, leaning my head back onto the wall in the hallway.

After hugging for a long time, Ethan and I had let our foreheads touch and we'd kissed each other very lightly before saying goodbye. We'd agreed to leave it a while before we spoke again, deciding that he would be the one to initiate communication when he was ready. We also agreed that over the following day, I would tell Naomi and Tim what had happened so that they knew. We'd figure the rest out as we went. I told him how much he meant to me. I knew we would never be able to go back to being such close friends as we once had been, and he knew that, too.

My whole body panged with sadness. It had been the right thing to do, but I was going to miss him.

"Maya." I heard Dad's voice coming from the sitting room.

I opened my eyes. My adrenaline started to pump again, exhausted from running its marathon around my body. I ambled slowly towards his voice and peered around the door frame. Dad was sat in his work suit, his reading glasses propped at the end of his nose, the newspaper on the sofa next to him.

"Shall we have a quick chat?"

I moved to sit in the armchair opposite him. I had barely any energy left. Through all of this, Dad's reaction was the one I was struggling the most to predict. It scared me. I put my hands in my lap. He took off his reading glasses and placed them on the table next to him.

"I've spoken to your mum," he began.

I looked at him, biting the inside of my lip, hard, almost piercing the skin.

"I know this must be very difficult for you, and I'm sorry about that."

He spoke clearly, shifting his eyes away from me to look down at his hands.

My throat felt sore.

"It might take some time for me to adjust to all this, but I want you to know that I am trying, and I will keep trying."

He looked at me, then, as if seeing me for the first time. He furrowed his furry brows and gazed at me longingly. I knew that a myriad of thoughts would be going through his head. But more than that, he would be reassessing who I was.

"I'm a bit behind your mum, but I am still right behind you, OK?"

He blinked and, slowly, his mouth almost turned into a smile.

"Dad," I said, my eyes raw. "Thank you."

I'd been so afraid, for months, about this day. The day that I'd have to think about my feelings, admit them to others, and face their reactions. I'd known, deep inside, that this string of confessions and discussions was inevitable. It had been emotional, exhausting, but it had also been empowering. I was lucky to have people like Ethan, like my parents.

I thought about Aisling in that moment, about how much I didn't know about her family and friends. How much did they know? And, if they did know, what had their reactions been like?

Part Four

Together

21

IN THE MIDDLE

Aisling

Edinburgh

January 2014

READING A BOOK. She was reading a book. Turning the page. Standing up in the library. She didn't see me, but I saw her.

At first, I glimpsed her hair. It was curly and loose down her back, like plastic ribbons which had been coiled with scissors, each individual strand shiny. Her jeans hugged her legs, and an oatmeal jumper hung like syrup off her body.

I'd come back to Edinburgh from Ireland straight after the New Year, which I'd spent with my parents at home, watching the TV and having dinner in silence. I didn't feel like doing anything, and I wasn't invited to much else anyhow. Karen had messaged asking if I'd wanted to come back to Edinburgh for Hogmanay, but I'd said I'd still be in Ireland. On New Year's Day, I'd flown back.

It was the second of January, and I'd come to the library to start the first essay of term, even though teaching didn't resume until the thirteenth. The essay was about *Paradise Lost*, by John Milton. I'd attempted to read it all over the holidays but failed miserably. It was far too long. It was making me reflect, though, on the Bible and the themes of sin and redemption from an academic standpoint rather than a personal one, which was somewhat refreshing. My essay was about the Adam and Eve narrative, and the relationship between sex and guilt, a topic on which I had at least a few things to say.

I stood there in the library with my backpack on, fresh from the cold. I'd just come to write my essay, and there Maya was, right in front of me.

I hadn't seen her in almost two weeks. I hadn't spoken to her. But I had thought about her, constantly. I'd mused about what she was doing, who she was with. I'd wondered whether she was thinking about me. But the truth was, I worried that she'd been doing the opposite. Was our next interaction going to be the same as last time, when we had barely spoken for weeks? Was she going to come back and suggest we just be friends?

The day I'd slept with Orla, I'd typed a message while I lay in bed that night.

> Maya, I miss you. Can we talk? I want this
> to end.

I deleted it quickly, cursing myself for being so ridiculous. She had wanted space; we had agreed not to speak. I couldn't just message her.

On Christmas Day, again, my hands had hovered over the phone.

> Happy Christmas Maya. Hope you're
> having a wonderful time at home.

I'd backspaced that as well, deleting the letters one by one, letting them disappear like smoke from a flame.

I imagined our conversations instead. Things like what she'd say about the poems I'd written over the holidays, or about my family and how they treated me, or about what happened with Orla if she found out. I wanted to be near her, to talk to her. Not being able to do that had made me fatigued and frustrated. I couldn't sleep, I wasn't eating, I couldn't focus properly.

Suddenly, having her there in front of me, that all felt so juvenile.

She obviously didn't feel the things I felt. She was back in Edinburgh, and she hadn't contacted me like she said she would.

I was sure she hadn't seen me standing there in the library. I backed away around the corner; then, when she was out of sight, I turned and marched swiftly out of the reading room.

I burst into the bathrooms, bashing open the white swinging door. I hobbled towards the sinks, took off my backpack, dumped it on the floor, and leaned my hands on the edge of the sink. I felt sick, like I might vomit. I looked up to the mirror.

The blue of my eyes was dull and gray.

Knots and grease ran through my hair.

My skin was pale and pimply like a lychee.

I leaned against the sink for a few moments, regaining my balance, letting the nausea fade into the distance, taking deep breaths.

The white swinging door to the bathrooms opened.

Oatmeal jumper.

Jeans.

Curly hair.

Dark brown eyes.

I stood upright.

Maya held the door open with one hand and stood completely still.

"What's happened to your ear?" she said, releasing her hand and walking into the bathroom. The door shut.

I touched the stitches. "Oh, nothing. I just ripped an earring out by mistake. Snagged it on my"—I reached for my sleeve—"clothes."

She stared at me, decoding my expression. "Is it painful? Are you alright?"

"Oh, yeah, it'll be grand, don't worry about me."

She paused another moment. It was clear she knew I wasn't telling the truth. "I didn't realize you were back. I thought you'd still be at home."

"No, no. I got back yesterday," I said. We ogled at each other for a moment longer. "You?" I crossed my arms, aware of my body, my face, my hair. I could feel the sweat oozing through my clothes.

"I got back a few days ago. Gabe and I went to Hogmanay together."

"Ah. You weren't with Ethan?"

Suddenly, Maya looked at me the way I wanted to look at her.

"We're not together anymore." She spoke softly. "I couldn't—" She paused. Swallowed. Touched her hair, moving it out of her face.

"You couldn't what?"

Maya walked towards me. Her body gravitating closer to mine; it drew me towards her as well. I took a step forward. Like a force. A magnetic field drawing me in. The space broke like a blank page receiving words, and we met in the middle.

We kissed.

She touched my face, my neck. I ran my hands through her hair, around the curve of her strong, beautiful neck. I kissed her and let the music play in my head. I let it be everything I imagined it would be. We twisted and turned, our lips dancing, our bodies rotating and swirling into each other. My eyes started to water. The tears ran down and down and down my cheeks. I hadn't cried in weeks, even though I'd desperately needed a release. I could feel her crying, too. Our tears, our bodies, merging into one another.

When she drew away from me, our faces stayed close, our breath meeting.

She laughed, wiping away the tears. I wiped them away, too. And laughed. It was a type of laugh I'd never felt in my body before. Real happiness. I kissed the salty droplets from her cheeks and planted them on her mouth. And there we stood, caressing in the girls' bathroom of Edinburgh library. And after all that time of not feeling I was enough, after all this longing, I was finally where I was meant to be.

22

LIKE A FIREWORK

Maya

I'll be right down! X

Aisling had messaged me. I smiled and put my phone back in my pocket as I waited outside the door to her accommodation, the January sun resting on my face. I shut my eyes and let it soak into my skin.

When I opened my eyes again, I watched people go in and out of the corner shop opposite her building. A man tying up his cargo bike and glancing over his shoulder at his children, who wiggled their pink and orange helmets around their heads and whipped their limbs around. "Stop that now," he moaned at them. A woman wearing glasses, tottering out of the door and pulling a shopper trolley behind her. A young boy, obviously a student, striding into the shop with his leather backpack and his polished beige brogues.

I felt so on top of the world that I wanted to go and hug them all, one by one.

Life was good. Aisling and I were going to get coffee and head to the library, where we would sit and get no work done at all while we got distracted by each other's presence.

It'd only been a few days since we'd kissed in the toilets of that very same library. Technically not the most romantic place ever, but it had been the most romantic thing that had ever happened to me. We'd laughed about it afterwards. The toilets of the library. Not that the place had mattered, anyway. Everything around us had seemed to fade away.

The last couple of days we'd spent almost all our time together; we couldn't help it and we didn't want to be anywhere else. We kissed, held each other, whispered into each other's ears, stared into each other's eyes, and went over the last few months in painstaking detail, telling each other our truest thoughts. Every time we touched, it felt like something from a film or a book, but also nearly inexpressible. I couldn't capture that feeling through language, that feeling of falling in love with someone who was also falling in love with you.

Even though we hadn't slept together yet, I knew it was going to happen soon. We both did. I would wake up in the night just thinking about it, my body shivering with excitement at the thought of being that close, that connected to Aisling.

After everything with Ethan, Christmas had been relatively quiet and melancholy. I had mainly stayed home with my parents, playing board games, watching TV by the fire, or doing some writing and drawing when I felt like it. Although I knew they were constantly thinking about my confession, my parents and I didn't discuss things further. It was just good to spend time together and to process things in one another's company. I could feel them, bit by bit, coming to terms with it all, and I was grateful for that.

I didn't feel like going to any of the meetups happening with people from school at bars or pubs over that Christmas period. I didn't want to put on a face anymore, to pretend to be someone I wasn't. After various events, people had messaged me to ask where I had been. I was sure they all knew Ethan and I had broken up. Perhaps they even knew why; news like that always spread rapidly. I'd messaged back telling them I had loads of family commitments, which seemed to work, and I tried not to worry about what they would inevitably be saying about me behind my back.

I had, however, gone to meet Naomi and Tim one frosty morning a couple of days after Christmas. We'd rendezvoused at a coffee shop near Notting Hill, exchanging hugs and saying how much we'd all missed one another this term. It was good to see them and hear how they were properly, rather than over a rushed text or a fleeting phone call. Naomi was having a great time in Brighton, where she was studying art. She'd just started seeing some guy and was going to his in Newcastle for New Year's to meet all his friends from home. Tim told us he had nothing to report and that his life was boring, to which we cooed in protest. He was working this year, living at his parents', and saving up for some traveling over the summer.

After a while, the subject of Ethan came up. They already knew, of course. As Ethan and I had agreed, I'd messaged them about the situation the day after Ethan and I had split up. As we sat in the café, though, I explained things to them further, and in the same way I had done with Ethan. I told them I was sorry; I knew this would affect our group and impact us all being able to hang out together. I was nervous as to how they'd react, but both were kind, saying they completely understood and that I'd done the right thing. I could tell, though, they were gutted. They knew, just as well as I did, that the four of us being together would never be the same again. Perhaps they thought I should have been more careful before jumping into things with Ethan, or perhaps I was just fearful that they thought that. I knew I would have to be the one to take a step back from the group, and I was sure they knew that, too. That fact went unspoken, but it saddened me more than anything.

I'd come back to Edinburgh to spend New Year's Eve with Gabe. We'd traveled together on the train from London. He'd asked me about everything while we sat there, the soft fields and glistening sea rushing past us. Finally, I admitted it all to him.

"It had been coming for a while," he said. "I suspected how you and Ash felt about each other. You did the right thing, Maya." His hand rested on mine across the train table.

Once back in Edinburgh, Aisling was on my mind the whole time. Before I'd seen her in the library, I thought about messaging her. In fact, I was spending most of my time considering in detail what I was going to say.

Then it'd happened, totally unplanned.

I'd spotted her out of the corner of my eye. I could feel someone watching me and when I saw the figure rushing away, I recognized her immediately. Her coat. Her backpack. Her long dark hair. I could tell even from the faintest movement. I'd instinctively followed her out into the corridor and then into the bathroom, swinging the door open to see her. My skin had risen into goose bumps. I'd felt so nervous, so unprepared, so caught off guard. But there had been something beautiful about that. The way we'd stumbled over words, the way we'd come towards each other like we couldn't wait a second longer. It was everything and nothing like I'd imagined it to be. It was perfect. Finally, I was with Aisling, and all was well. Even the insect had been hibernating that week; it was nowhere to be seen or felt.

There was one thing troubling me, though.

Aisling's ear.

I could see it had obviously been stitched. It bothered me just like the scar on her hand bothered me. She had told me she'd snagged an earring on her clothes, and the scar on her palm had just been brushed off as an accident, but I knew neither of those explanations were true. I knew it was linked to something she wasn't telling me. I'd suspected about her family, and I'd even tried to ask her about them, but she'd shrugged it off.

One day that week after we'd kissed, we sat in George Square Gardens. I'd brought two coffees to the library, one for her, one for me, and so we'd decided to take a long break together. She'd asked me all about my parents' reaction when I'd spoken to them over Christmas, and also about what happened with Ethan. I'd told her, briefly, and after I finished, I realized that it could be a good moment.

"Aisling, can I ask you something?"

"Ask away," she'd flirted.

"What happened to your ear, and your hand?"

There was a long silence. OK, maybe it hadn't been a good idea to ask, I thought; maybe I've overstepped the mark. I got uncomfortable and peered down into my lap.

"I'm sorry, it's OK if you don't want to talk about it now, obviously, I just— I'm only asking because I care about you, and when you're ready to talk to me, I'm here."

I had turned to face her, and she did the same. Aisling glanced at me in a way she never had before. She knew what I suspected. It was as if she was trying to calculate how I'd come to this conclusion, and for how long I'd known.

"Thanks." Her look melted into something gentler.

There was a pause.

"I have trouble with my family sometimes."

"Do they hurt you?" I'd asked her after a moment.

Aisling had looked directly at me, and then, briefly, she scanned the square to check no one was nearby. There was no one around. She didn't nod and she didn't shake her head. Her eyes didn't even blink.

"Sometimes."

I reached my hand over to hers. As I held it, it felt so cold. "And they still do? Hurt you, I mean."

I could see her trying to think where to start, what to say, how to say it.

"I'm sorry. We don't have to talk about this if you don't want."

"I want to talk about it sometime, I'm just, I'm not—"

"Of course, whenever."

I had felt completely sick at the thought of something bad happening to her. My stomach had twisted, draining itself of every inch of air and liquid, turning into solid rock. I wanted to absorb Aisling into myself, to encase her in my own body. Instead, I just clung onto her hand, desperately, then leaned over and kissed her. All I wanted was to be close to her, to be near her.

As I stood outside her building that morning, waiting, I replayed this conversation.

Then the door to Aisling's building clicked open behind me.

Aisling stood at the top of the stone steps, her pale skin radiating like the sun, her blue eyes like droplets of water, her cherry-red lips wrinkled together, about to burst, like a firework.

"Alright, you."

She trotted down the steps towards me.

"Fancy seeing you here. It's been a while."

She planted the softest kiss on my lips, holding my cheek in her delicate hand. It made my insides melt, smelling her fresh, powdery skin, tasting her minty toothpaste.

"I know, it's been at least nine hours," I muttered.

She giggled and kissed me again.

We let our fingers slowly interlink and walked to the library through the meadows. Some people stared at us or gave us funny looks, but we didn't care. We swung around each other, connected always by the touch of our skin.

23

LIKE ONE

Maya

A FEW DAYS later, Aisling and I sat in my kitchen at about nine p.m. She was peeling a clementine, and little droplets of juice sprayed from its skin, making the room smell of orange zest. As she undressed the fruit, she laid the peel out on the table in front of her, making the shape of a smiley face. I sat on the countertop, waiting for the kettle to come to the boil.

"But seriously, who knew I was so artistically talented. You've got competition," Aisling said, pointing at the smiley face and looking up at me. I shook my head and laughed, jumping onto the floor.

As the kettle rumbled, I looked over to her. She was popping a segment of the orange into her mouth and admiring her orange-peel artwork. I poured the boiling-hot water into the two white mugs, then took them over to the kitchen table, putting them down and sitting opposite her. We started to chat and laugh about the evening we'd just had.

We'd been to the first poetry society meeting of term; it

started back a bit before lectures began. To celebrate our return, we'd also gone out for pizza and drinks all together afterwards.

It felt good to let people see Aisling and me together, but I could tell it also made us both slightly uneasy. Things were still so new, and we were uncertain about how to navigate it all. Before entering, we'd stood outside the meeting room under some offensively bright LED lights. We held each other's hands and confirmed what we'd agreed. We had decided to simply let people see that we were together; they would figure it out. We weren't going to make a big deal of it or anything like that.

As soon as we pushed the door open, I could see Harry's face. Like a meerkat, he poked his nose towards us, assessing our joined hands. Aisling and I sat down at the back together. I felt so anxious, but the movements of her body in my peripheral vision made me feel like it would all be OK, just like it had when I'd sensed her presence in the library. Everything she did made me feel better. The way she unpacked her books and spread them out on the table. The way she got her water bottle out from her bag and took a sip. The gentle hush of her breathing. I looked around the room. People were acting normal; they weren't looking at us unusually or anything like that. As more people entered, they started to greet us, tapping our shoulders, waving at us across the room. Aisling and I held hands under the table, linking our fingers together, letting the sweat from our palms mingle.

After the meeting, Ruby asked if we were together now, and, blushing, hot-faced, we'd confirmed that we were. Everyone, except for Harry, had been excited and, apparently, unsurprised.

"I just knew it," Isla had said. "I could just tell there was something between you two. I'm so happy for you guys." She'd grinned.

"Oh, me too, me too," flapped Ruby. "I so called it. Didn't I say to you there was a vibe between Aisling and Maya, Rohan? I totally said there was a vibe there."

Rohan had nodded. "She did say to me, and I told her she was no genius because it was completely obvious."

We all laughed on our way to dinner, mine and Aisling's hands still entwined, but no longer sweating. We gripped onto each other tightly, feeling utterly relieved.

As Aisling and I sat in my kitchen drinking our tea later that night, there was a pause once we'd finished dissecting everyone's reactions. We held eye contact, and I bit the corner of my lip. Under the table, we felt for each other's legs and our shins rubbed together. Her touch made all the muscles in my body weaken. I put down the mug of tea, then raised myself over towards her. Our lips touched. Hers were wet, and I could taste the orange on her tongue. She bit my lip gently as we pulled apart.

I sat back down and felt myself getting nervous. I was warm and the pit of my stomach felt tender.

"Is Gabe here?" Aisling whispered.

I shook my head, wetting my bottom lip with my teeth. "He's on a date."

As we sat there, I could feel the shallowness of my heartbeat near the top of my chest and in the base of my throat. I moved slowly. I stood, took my mug of tea, and put it in on the countertop. It was still full and hot, but I didn't want to drink it anymore. I turned back to look at Aisling. She was staring at me, leaning back in her chair, a slight smile on her face. I picked up her mug, moving that away as well, then went back towards her as she also stood up. We were close to each other. Our shallow breathing aligning.

We kissed again. I could feel her hands on my jawline. She guided me so I was perched on the side of the table, and she ran her fingers down my spine, then drew her face back so our noses and foreheads were touching.

"Do you want to?"

"Yes," I breathed.

• • •

AISLING LOOKED DEEP into my eyes as we lay on my bed.

Her body was on top of mine. Her skin touched mine. It was soft, then rough at the ends of her fingers, behind her knuckles, and where the scar was on her hand.

I looked at her. Just her. The color and complexity of her eyes. Like a forest of bluebells at the beginning of spring. Each streak of her hair, falling in front of her face, like a perfect black frame, luxuriating over her breasts. I ran my hand down her back; I felt the bumps, the wrinkles from the scars there, too. I kissed her nose, like a rolling hill flicking into the horizon. The moisture on her red lips which curved like the petal of a flower.

The rhythm of her body made my body beat.

Her fingers, rubbing me.

Inside me.

My breath, heaving. My chest, sweating.

Heat.

Gently kissing my lips, then my neck.

Down my chest, my breasts.

Down.

Down my stomach.

Tingling.

Like a flock of birds taking off inside my body. I was lifted upwards.

Downwards.

Torn in half.

Exploded.

Aisling sucked me, licked me, kissed me.

Fast, then slow, then steady.

Steady.

Like a ship,

rocking,

side to side.

My skin felt like butter, melting off my body.

I shivered and felt myself giving off steam.

I had never felt like this before. She knew what to do with me, and somehow, I knew how to respond. It was a conversation I never knew I could have. One I wanted to have again, and again, again, and again.

We lay beside each other. Panting. Our stomachs rising and falling with each other. Our figures must have looked so different. Different skin, different marks, different hair, different shapes, different lengths.

But we felt like one.

As we started to drift off, Aisling whispered to me, like an angel: "I've never fallen asleep next to someone before."

In the darkness, I clutched onto her hand. Our skin was woven together. We both felt safe; we trusted the other to protect our own unconscious body. I held onto her, tight, a smile spreading across my face, my whole body still tingling.

24

BOOKSHOP

Aisling

FEBRUARY 2014

I SAT IN a café on Grassmarket a few weeks later, early in the morning. I was the only customer in there. I'd paid for a coffee, which was sat, half drunk, on the oak table in front of me. The froth had melted into small bubbles which popped into the brown, milky residue at the bottom of the light blue mug.

I'd kept my coat on. It was freezing in there. The door had been left open, and outside the frost had melted into a mellow but cold wind. I had my notebook in front of me and was writing down words and phrases.

"Would you like anything else?" said the woman working there. She wiped the table next to me with a cloth.

"You're grand. Cheers, though."

I glanced up at her and offered a shallow smile, then continued to write, checking my watch. It was 8:56 a.m. I picked up the mug and swallowed the rest of the coffee. Then I shut my notebook and stared out the front window of the café, my heavy boots jiggling around as I tapped my foot nervously under the table. I waited for minutes, watching every person who walked past, looking for familiar pieces of clothing, for her curly hair and incandescent stride.

I turned to see if the woman was still behind the counter. My mouth felt dry.

"Actually, could I grab a water, please?"

The woman gave a subtle, almost imperceptible eye roll, and a couple of moments later she placed a full glass in front of me. I took a few gulps, then licked the residue off my lips.

"Cheers."

Getting up to leave, I put a tip on the table, then stuffed the notebook and pen into my backpack. I decided to go and have a look. Perhaps she'd come from a different direction, and I hadn't seen her.

The last couple of weeks had been incredible. Being with Maya was everything I'd hoped for and everything I'd wanted. It felt so natural and yet so exciting. I woke up every morning looking forward to the day, and I felt a pulse running through my body at all times.

We were inseparable; we stayed with each other most nights, except when we needed to catch up on sleep. As we lay in bed in the evenings, we'd read each other poems or extracts of books, and we'd just close our eyes to the sound of each other's voice, commenting on certain words or lines. We worked together, got lunch and dinner together, often with Gabe, and we went to poetry society and the socials together. Now that everyone knew about us, things felt easy. I'd never felt this sense of openness. I didn't have to be afraid; my parents, family friends, people from school, they were all so far away. We could hold hands as we walked down the street or kiss in the middle of a park if we wanted. It was mad.

Sleeping with Maya had been a holy experience. We talked and communicated with our bodies and our eyes and our voices, reading and sensing what the other needed and wanted.

The closeness of being alone with her meant that when we were around others, we had access to a door to which only we had the key. We had started going to this late-night spoken word event with people from society, and some of them even performed there each week. Sometimes we'd hold hands under the table and sometimes we'd end up on opposite sides of the room. Regardless of where we were, we would be on each other's radar.

But there was also a vulnerability in being with Maya that I'd never had with Orla. Orla had known about my family, at least roughly, but we'd never spoken about it explicitly. With Maya, things seemed to be blurring. She'd asked me about my ear and my hand soon after we'd first kissed. We'd been having a coffee in George Square Gardens at the time. The conversation had been jarring for me. I hadn't felt like it was the right time for her to bring it up, and I wasn't ready to talk about it.

The truth was, at that point, I didn't think I'd ever be ready to talk about it. But after we'd first slept together later that week, I knew she'd seen and felt the scars on the rest of my body, the ones which were normally hidden. Of course I realized it was inevitable. Orla had never commented on those, but I could see in Maya's eyes that she'd wanted to ask about them. As we lay there afterwards, I knew she was thinking about them. She didn't ask, though. She just stared deep into my eyes as she ran her hands over them; then she'd kissed my nose, cheek, mouth.

I'd left early in the morning after that first night together. I felt so exposed. I knew I was being perhaps too protective of myself, but that was all I'd ever known. I'd walked home through the early-morning mist, the crack of the sun breaking the sky. I sat on the steps of my accommodation block until people started to bustle through the streets. I watched the world coming alive, awaking from slumber.

It was overwhelming for me, thinking about how to express

my past and what it would do to our relationship. Would she run away, abandon me, just as Orla had?

I'd gone to find Maya in the library a couple of hours later, knowing deep down that if I momentarily retracted, I would always incline back to her.

Since then, I had given in to this vulnerability, even if I hadn't said as much. Strangely, this had only made me feel closer to her. She hadn't asked again, but I knew that I trusted Maya. I knew she would wait for me to talk about it, and I knew she would listen. Even if she didn't understand, I knew she would try.

It didn't matter that this would bring back memories that I'd been able to get away from in Edinburgh. Memories of Mother. Her disdainful words to me, the feeling of her skin on my skin, and the sounds of my own unreceived cries for help. I wouldn't enjoy it, but I was ready. I wanted to share this, I wanted Maya to know these parts of me, even if just a little piece at a time.

I was waiting in the coffee shop on Grassmarket that morning because I wanted to surprise her. She had told me she hated surprises, but I figured this was only a small one. Gabe had shared with me a couple of days ago that on Wednesdays Maya had a lecture at ten a.m. and she usually liked to go to the bookshop before, around nine a.m., to have a browse.

I walked down the street from the coffee shop and into the bookshop.

My eyes darted around, trying to spot her, to feel the comfort and the warmth of her presence near me. I wandered about, picking up random books, feeling their rough covers in my hands, clearing the dryness in my throat. I trotted downstairs to the poetry section, but it was empty. Turning around, I went back up the stairs, and as I emerged at the top, there she was.

Maya stood at the back of the shop, reading the inside cover of a red book. She wore a long, dark green coat, pink gloves, a beanie, and her backpack. Curls protruded out the bottom of the hat. I stood there and watched her. Every time I saw her, I still felt my heart overtaking my body. I observed her eyes moving across the words, just like I had done in the library that day before we first kissed. Her tongue on her bottom lip. Her foot propped out to reveal the inner part of her leg. I looked at her as she put the book back on the shelf, pressed the spine into the gap, and turned around. As she saw me, her mouth opened.

"What are you doing here?" She grinned.

"Complete coincidence," I said, feigning nonchalance. I picked up a book and pretended to read the back cover before replacing it.

She giggled and moved closer to me.

"Oh, really? Gabe didn't happen to tell you my Wednesday morning routine or anything like that then?"

"Ah, he might have mentioned something about it."

I smiled and fidgeted my feet. Maya walked towards me, maneuvering me around into the corner of the shop, where no one else could see us. She quickly glimpsed around, and then kissed me up against the shelves, her pink gloves caressing my face.

"So, how about a coffee before your lecture then?" I murmured to her.

• • •

LATER THAT EVENING, I stood in the kitchen of Maya's flat with her and Gabe. The sound of laughter seeped through the air. Gabe was doing an impression of one of his lecturers. His impersonations were unbelievable; we constantly told him he should be an actor. Sometimes, when we'd watch the TV together, he'd pretend to be the politicians on the news or certain celebrities.

Whenever he did, Maya and I would be in stitches, unable to control our giggles.

As he finished, Maya and I wiped our eyes and continued what we had been doing. I was, slowly, learning how to cook through Maya's guidance. At that moment, I was cutting up some carrots, and she was pouring some oil into a pan. Another pot, full of rice, was bubbling away at the back of the stove.

Gabe jumped up to sit on the kitchen countertop and started to fiddle with a bottle opener.

"I saw one of those guys from poetry society in town today," he said.

"Which one?" Maya asked, scooping some chopped onions into the frying pan.

"The mousy one who runs the show."

"Harry?" I asked, laughing a bit at this description.

"Yeah, yeah, Harry."

"Christ, he's not happy I'm going out with you, Maya," I admitted. "Gives me the dirtiest looks. He has a real thing for you."

I turned to Maya, who was shaking her head as Gabe continued.

"Isla was telling me the other day that—"

"Isla?" Maya butted in, putting down her knife and turning to Gabe, her eyes narrowing. "What're you doing speaking to Isla?"

I stopped and spun around to look at Gabe, too. His cheeks were red, and he pushed his cloudy glasses nervously up onto his nose.

"Um. Well."

Maya and I exchanged a short look.

"Is that who you've been dating? Oh my gosh, I knew it. I just knew it." She threw her hands up in the air.

"Well, we met at your party, and we've sort of been seeing each other since then, yeah." Gabe shrugged, trying to suppress his smile but doing a bad job of it. "I didn't tell you because I didn't want you to get all excited like you're doing now. I didn't want to put any pressure on it."

"Call me matchmaker!" Maya beamed.

"That makes sense," I reassured Gabe. "And that's great you've been seeing each other, Isla is really lovely."

"Thanks, Ash." Gabe nodded, smiling properly and trying to ignore Maya. "I think so. I really—I really do like her."

Maya stopped what she was doing and rushed over to give Gabe a big hug. He hopped down off the counter, and they embraced as he told her repeatedly not to make a big fuss, please. I caught Gabe's eye over her shoulder, and we exchanged a warm grin.

That night, Maya and I lay in bed facing each other in the dark. We whispered to one another. She told me she was so happy about Gabe and Isla, joking that she could convince them they owed her endless drinks now given they'd met through her. My stomach scrunched, but I smiled softly, letting a breath out my nose. I joked back that maybe we could even get Gabe into poetry now.

Maya reached over and touched my cheek, then ran her hand over my shoulder, down my arm, and grasped my hand in hers. She felt the scar on my palm with her thumb, running her touch over the crease lightly.

"Maya?"

"Yes?"

"I'm ready."

Immediately, she knew what I meant.

She linked her fingers with mine and squeezed my hand tight. I could feel her nodding, even though I couldn't make out more than her outline in the darkness.

"OK. There's no pressure, but if you are ready, then . . ."

I swallowed and shut my eyes, rolling onto my back. Our hands interwoven.

I told her I wouldn't go into the details, but I wanted to share some of it with her.

I told Maya that my mother was, and had been through my whole childhood, an alcoholic. I told her that she had beaten and abused me as a child whenever we'd been alone, whenever she'd been drinking and sometimes when she hadn't been. I told her that I was certain my father had always known about this and had done nothing to stop it. I told her my siblings had also done nothing, even though I was sure they knew. I told her I'd never understood why I had been singled out; that question haunted me every day. I told her that my whole family feared my mother and that her manipulating and terrorizing had silenced us all. I told her that was what my scars were from, on my back, my stomach, my ear, my hand. And then I stopped talking, my eyes wrinkled tightly shut.

I breathed in, I breathed out. I could feel her next to me, her body expanding and shrinking.

I felt so vulnerable. As if my blood and guts were on the outside. Like I'd stripped myself of my own flesh, my barrier, my covering. I'd never, ever talked to anyone about my childhood before. Not even Orla. But I wanted Maya to know. It'd come out in broken sentences and slow snippets, but it had come. It'd made me feel bare, but it was over. I didn't have to talk about it again. I didn't have to relive it.

I reached up to my face with my free hand in the darkness, and I felt tears running down my cheeks. I was crying.

And then I felt something else. I felt Maya. I felt her embracing me. The softness of her arms, her fingers, her legs, her lips. Her steadiness. Her body around mine. I felt her touch, her comfort, her warmth, her protection. I felt her kissing the tears on my cheeks, running her hands over my back and through my hair. I felt the air of her voice.

She thanked me for trusting her. Then she paused, and I could hear her beginning to cry as well.

"I am so sorry you went through that. You don't deserve to have been through anything bad, Aisling."

At her words, I turned over onto my side so I could hold her properly. I felt the tears still slipping down my skin, my body shivering. And I clasped onto this moment, this reality, trying desperately not to slip back to the other one.

I felt her lips next to my ear.

"Ash, I love you. I love you."

Our bodies tangled, our hearts beat together, our breathing synced. I'd never felt so close to anyone in my life. Telling her all that had felt like giving her a part of myself. It felt terrifying, completely alien, to feel someone love you and to love them back. It was so overpowering, almost maddening. The feeling so strong and new. It made me cry even more.

"I love you, too, Maya," I whispered back to her, moving closer still.

25

YOUNG LOVE

Maya

MARCH 2014

ABOUT A MONTH later, spring was starting to tumble into the air and press up through the earth. The bite of the cold was fading, and the darkness was waning.

Edinburgh was as beautiful as ever, but it wore a different costume. Sunshine burst through; flowers started to unfurl. The rain-covered, lamp-lit glow of the dark, tall buildings and the cobbled streets was replaced by light bouncing off the castle walls and the glimmers of sun on the tantalizing sea.

At around ten p.m. one Wednesday night, I unlocked the door to my flat with one hand and pushed it open. Aisling had her back pressed up against it. She was kissing me. My other hand was resting beneath her ear. As we entered the flat, our lips still locked together, we rotated so I could flick the door shut with my foot. We walked together, me backwards, her forwards, in perfect step towards the sofa, collapsing our intertwined bodies down onto its pillows.

I paused and drew back from her, pressing my finger to her lips softly and temporarily.

"Gabe?" I called.

There was no answer.

We fell into each other again, our bodies rolling over each other on the sofa. I felt bad sometimes for Gabe that we always came to our apartment, but it made sense given I had a double bed and Aisling had a creaky single. Avoiding the times of the day that Toby was likely to be cooking, we had been over to hers occasionally to have dinner with her housemate Karen, who I'd come to like. Karen had even come to poetry society drinks a few times as she knew Harry from home. Even though she was a self-confessed poem-loather, she was always "great craic," as Ash would say.

That evening, we'd been to a reading at a bookshop near my place. The book was a literary novel which had just come out, and the author had read out a few extracts of it before being interviewed by another writer. Aisling had suggested that we go because she thought it might be interesting. It had ended up being fairly stimulating. Aisling had even asked a question at the end.

There'd also been free wine at the event; I'd had a glass. I could taste it on my tongue as Aisling and I kissed. Mostly, I tried not to drink as much anymore because of what Aisling had told me about her mother. We hadn't discussed it or anything, me cutting down. I knew she'd tell me not to worry and to drink whenever and whatever I wanted, but I was following my own instinct and I knew she'd noticed.

When she'd told me everything about her family, I had felt sick as she spoke, my body like a hollow corpse only housing the insect. I wanted to wrap Aisling up and never let her go. I felt so close to her, so trusted, so in love with her. I never wanted anything bad to happen to her and I only hoped I could be enough.

"Alright, lovebirds," Gabe's voice boomed through the room.

Aisling and I sat up in shock like puppets being pulled with strings, our noses and lips hot.

"I thought you weren't here. I called your name when we—"

"Sorry, had my headphones in." He grinned, releasing a chuckle.

Aisling and I straightened out our clothing. She reached her hand over to cup mine as we sat there, smiling angelically at Gabe.

"How was the reading?" he asked.

"Fascinating, actually." I nodded.

"Oh, yeah? Are you shifting your loyalty from poetry to novels then?"

Aisling let out a breath through her nose and scrunched up her face.

"Do we have to choose?"

I watched her as we sat there and felt myself relax. Aisling had that effect on me; being with her made me feel at ease. The insect had mostly crumbled into nothingness over the previous weeks. I'd even told Aisling about it one night as we lay in bed, and it felt like a relief to speak of it. I'd never named it out loud before, only in my thoughts. Somehow, when I explained it, it didn't seem so scary.

With Ash, I existed without the facade I'd spent so long formulating. This authenticity had seeped into how I felt about other things. My hair, my clothes, my demeanor. Our love didn't feel dependent on those things.

"Right," Gabe said, clapping his hands and standing up. "I'm headed out to Isla's, so you kids be good." He pointed his finger at us, his head tipped forward.

"Can't promise," Ash muttered.

"Say hi to Isla from us." I smiled.

"I will."

"Tell her we're still waiting for our double date," I added. Aisling gently covered my mouth with her hand and told me to let poor Gabe go. Gabe smiled at me, pressing his eyelids down in fondness.

As Gabe was leaving, he twisted to us and looked between our faces. He pushed his glasses further onto his nose.

"Young love," he hummed, touching his chest before blowing us a kiss and closing the front door behind him. We watched it shut, and, turning back to each other, our bodies tumbled together again.

26

EASTER

Aisling

County Clare

APRIL 2014

"TIME TO HEAD to Mass everyone!" called Ma from the bottom of the stairs.

It was Easter Sunday. I'd come home just for a few days. We didn't really get proper Easter holidays at Edinburgh because revision and exams fell right at the end of the second semester, but some people did take a few days out to go home, see their loved ones, or maybe even revise from their family home. Of course, this was not something I shared with Ma when she called up to insist I come back for at least the Easter weekend. I didn't want to return to the house of my parents.

"You must come back," she'd instructed me on the phone. "For the love of God, it's Easter, Aisling. The time of Christ's resurrection, remember?"

"It's just, I've got so much work and my exams are coming up."

Silence on the other end of the line.

"OK," I'd said, "I'll come."

After we'd spoken on the phone, I'd sat on Maya's bed with

her next to me. I slipped my hand into hers, feeling our fingers link together.

"I don't want to go. I want to be here, with you," I muttered.

I ran my hand over her bedsheets, which were covered with blossom patterns, creased from us rolling around and sleeping curled up next to each other.

"Don't go," she whispered, gently running her hand up and down my arm, then through my hair. "You don't have to."

Thoughts ran through my head like passing trains. Then I was hit by them. Perhaps something worse would happen next time if I didn't go home this time. Perhaps Ma would punish me more because I hadn't cooperated.

"You won't be here anyway, though, will you?"

I looked into Maya's dark eyes as she shook her head.

"I'll be going home for a week or so. But you could come with me?"

I thought about spending time in Maya's parents' house, them cooking me dinner, asking me questions about myself.

"Are you sure?"

She nodded.

"That would be great."

I touched my hand to her soft face. Maya leant her head into it, so it was between her cheek and shoulder. She closed her eyes. As I looked at her, my thoughts went back to my own parents. The fear of retribution rematerialized.

"I should go home. Just for a few days."

Maya sat up straight. "Really?"

"I think I should."

Maya bit her bottom lip. "Ash, are you sure? I don't want you to . . ."

I nodded.

"OK," she'd conceded. "But you can always come to mine

whenever you want. I can give you the address, and you can just come there if you need to."

Back home in County Clare, as I got ready for Easter Mass, I thought about this conversation. The memory of it relaxed me. In my purse I had her address written down. I'd folded the small piece of paper over four times. It was in her handwriting. Even that was comforting. I thought about where she was at that moment. Probably in her stripy pajamas, at the kitchen table, having a coffee and reading the newspaper with her parents as they all smiled and laughed together.

I got into the Citroën with the rest of the family, my parents and all three siblings, who had also come home just for Easter weekend. As we drove to church through the winding lanes and the emerald hedges, I wished I was back there with her, on that bed in Edinburgh, feeling her skin and lips touching mine.

"Aisling!" Ma's voice was raised over the vibration of the tires on the road.

"Yes?" I answered, coming out from my daze.

"I just asked you a question, for God's sake."

I looked around, and everyone's eyes were on me. Even Pa, who was driving, was glancing at me in the rearview mirror.

"Sorry," I mumbled. "What was it?"

"Did you take Mary's blue sweater? The one with the thin stripes. She's lost it."

"Oh," I responded. "I thought that she'd given it to me. It was in the pile of stuff you didn't want over Christmas?" I turned directly to Mary.

"No," Mary said, being short with me. "It wasn't, but it's alright. Just keep it."

"You must give it back." My mother's voice was loud and harsh.

"Course I'll give it back. I haven't brought it with me, though, it's in—"

Ma made a sudden sound, which grated in the back of her throat. "Ugh, of course." She turned to me over her shoulder from the passenger seat. "Very crafty."

"I didn't know, I—"

"It's fine," said Mary, cutting me off, eyeing Ma. "Don't worry."

Normally, the presence of my siblings made me feel safer, protected. I had always thought that Ma wouldn't do anything to me if they were around. When they visited, it was usually only at night, alone in my room, that I felt scared. But this time, even with my siblings there, I felt on edge. Something was looming.

We turned into the church car park, and my stomach twisted.

"Out we get," said Pa, pulling up the hand brake. As we opened the car doors, the faint sound of the organ seeped out from the cracks of the building.

• • •

WHEN WE GOT home from Mass, we sat down for lunch. The smell of the lamb dripped in the air, the mint sauce, the daffodils which were bunched up in a vase at the middle of the table. Everything smelled fresh and calm, but the air felt stiff.

I looked towards Ma as I sat there. She was rearranging the bowls of slimy potatoes and watery broccoli on the table, straightening them out and aligning them. Pa came into the room carrying the plate with the lamb balanced on top. A big hunk of flesh ready to be carved. Ma clapped.

"Ahhhh," she sighed. "Gorgeous."

For the last couple of months, I'd been eating vegetarian with Maya. I hadn't told Ma and Pa, mainly because I never spoke to them, but also because they'd resent it. They'd think it was an attempt to better myself, when the truth was, I simply preferred it. Plus, if anything was going to turn you off meat, Toby's meals would do it.

After so many months without, the sight and smell of carcass made me feel queasy.

We said the prayer. I could feel Sean next to me, clenching his hands together super tight.

"Aisling," said Ma, as if bored, unclasping her hands, "you're youngest. Pass me your plate and we'll get you some lamb."

I watched Ma stand up and start to slice the meat with a sharp knife and carving fork.

I hesitated, feeling everyone's eyes on me.

"I'm alright without, thanks."

She looked at me, sternly, her eyebrows furrowed, and then she dropped the knife. It clanged on the plate, making everyone and everything jump.

"You're not a feckin' anorexic now, are you?"

"No," I said, avoiding her eye contact. "It's just, I'm sort of not eating meat any longer."

She raised her voice. "Why on earth would you do that?"

I shrugged. Her body collapsed into her chair. I became very aware of the knife lying just in front of her on the table.

"Well, Jesus Christ," she hissed, picking the knife back up mindlessly. "Why didn't you tell me? Stupid girl."

"I—"

She stood up again and stared at me, muttering under her breath.

It was as if she were casting a spell, chanting a curse, and everything else was melting away. I was alone with her.

"You are so ungrateful. We've spent money on this, time cooking it, and you won't even sit here and just—"

"I'll eat it then," I interrupted. "I will, I don't want to—"

"You can't even just sit there and eat with us like a feckin' normal person; you must make this about you. You're selfish, a selfish little child, that's what you are."

Her voice was getting even louder.

"I'm not trying to be selfish," I said, attempting to stay calm. "I just don't—"

"You don't think about anyone but yourself. That's your problem," she cut in. Her free palm slammed flat on the table. Her whole body was starting to shudder. Her arms, her stomach, her face.

I shook my head and shut my eyes. "Don't say that."

"Well, that's selfish. You are just so absorbed in your own—"

"Stop it, stop it," I said, raising my own voice, opening my eyes.

"I will NOT stop."

Slowly, her nostrils flared, she continued to speak as she moved around the table towards me.

"It is TRUE."

"It's not," I muttered, shutting my eyes again, shaking my head still, willing her to be wrong.

I could hear other people's voices around us, but I was only aware of her in the room. When she spoke again, I knew she was close.

"There are rules, Aisling. Things must be done in certain ways. You can't just decide the way things are and go about your life with NO consequences."

I opened my eyes.

She was moving past the blur of other voices, figures, objects, and she was waving the knife in front of her, rolling it around as she spoke.

"You ruin EVERYTHING."

"That's not—"

I couldn't take my eyes off the knife, but I also couldn't seem to move. My body was motionless. It was as if I were a little girl again. That little girl staring wide-eyed at her mother, paralyzed by fear but also wanting love.

Everything seemed the same now. The next few moments felt inevitable, and all I could do was shut my eyes again, tight. I did.

I could hear noises. I could hear people saying her name, but I couldn't see the bodies attached to the voices. I could feel her getting closer.

"For FUCK'S sake, Aisling!" Ma yelled.

Then there was another yell from a different voice, a scraping sound, and a loud blast of noise.

Did I feel pain? Was that the noise of something dropping on my foot or hand or leg? Had it already happened?

Slowly I opened my eyes and then drew in a sharp breath.

The other yell had been from Sean.

Sean was shouting, pushing his chair back, standing right in front of me.

"BACK THE FUCK OFF! DON'T TOUCH HER," Sean was screaming.

For a moment, I had forgotten that Ma and I weren't the only two people in the room.

Everyone's heads were turned to Sean, who now stood between Ma and me. Everyone's breathing was heavy. Ma's, Sean's, Pa's, Jack's, Mary's. Their chests and rib cages flapped like sails in the wind. Their chairs were all drawn slightly away from the table, their eyes all wide.

"JESUS CHRIST, IT'S ONLY SOME LAMB," he shouted, moving his face closer to hers.

She was still holding the knife, but his reaction had stopped her in her tracks.

"Let her do what she wants," he instructed, his voice firm, but quieter then.

Sean waited a beat and walked out the room.

"Where are you going?" I asked, my voice trembling involuntarily.

He turned in the doorway. I could see, now, that he was shaking.

"C'mon. I'm taking you away from here. We'll—" He swallowed. "We'll go to the shop to get you something else to eat."

He paused and stared directly at Mother.

"Everyone needs to calm down, or something bad is going to happen here."

There was a long, painful silence.

Stunned by his behavior, I stared at him, trying to figure out what was going through his mind. I looked at Ma. She was gaping at Sean, her expression livid, but her mind faltering, her words and fury failing her. He was a strapping lad, and with him next to me, she couldn't touch me. She wouldn't. His anger was palpable, and his veiled threat hung in the air.

"Ash, c'mon."

I got up slowly, feeling all eyes still on Sean. I followed him out into the corridor and we put on our coats before walking out the front door.

• • •

AS WE DROVE to the shops, I started to feel myself steady. It was a relief to get out the house and realize there was a whole world out there.

I could hear Sean sniffing. I didn't look over in case he was crying. This was the brother who had left home when I had been just ten. I knew he had a job of some sort, but I didn't know what it was. I knew he was living with mates in Dublin, but I didn't know who they were. We weren't close, we never had been.

"Why did you do that for me?" I whispered after we'd driven about ten minutes.

Sean stayed silent. I turned to him, and he clenched his jaw. His eyes were wet. His hands gripped the steering wheel, tight, but still jittered ever so slightly.

Eventually, we arrived at the shops. Sean pulled up at the side of the road. The engine rumbled and then he flicked it off and pulled up the hand brake. We sat in the quiet. The streets and pavements were empty, apart from the odd person. Most people were at home, sat around the table with their family, laughing and joking, eating, drinking.

"I'm sorry, Aisling."

When I looked over at him, his eyes were bloodshot.

"I should have kept more of an eye on you. I've been such a coward. I didn't let myself— I must've been too scared, or stupid, or something."

"What?"

Sean slumped back into his seat, rubbing his hand on his forehead. He looked out the window.

"I don't know why it took me so long to realize how much I've abandoned you. How much Jack and Mary have abandoned you, too. How much you've been through. I don't even know why I chose not to see all of this was happening. I mean, like, I saw it. I saw the injuries, but I didn't connect the dots. It's like I was wearing a blindfold or being tricked into believing another thing had happened. That you'd just slipped or scraped yourself by accident or snagged your earring on something. And today, Christ, what would it have been today? That her hand had slipped, that she dropped it by mistake or—" He stopped talking, not wanting to think of it. "It sounds so stupid now even. I'm just— I'm so sorry."

He paused.

"Can you forgive me?"

I didn't know what had happened to make Sean realize the truth, but I'd never felt more grateful.

"Thank you, I forgive you. It's alright," I said to him.

"It's not alright. I haven't–"

"Sean, stop. There's no point."

He looked into my eyes, leaning his head against the headrest. Sean put his hand on mine across the car.

"So, do you have someone? In Edinburgh?"

I considered this question. I wanted to inquire why he was asking about this now, but I didn't have the energy just then.

"Maya. She's called Maya."

He nodded and smiled at me.

We sat there while the windows of the car got less foggy, the sights around us becoming clearer. I thought about how this must be what it feels like to be happy around your family. Thinking about it, sitting there, feeling safe almost, I bit my lip as my eyes started to fill with water.

Sean put his arms around me, and I leant into his shoulder.

At the corner shop, I got some vegetarian quiche, which cost over five euros; then Sean drove us back to the house.

When we pulled up, he unbuckled his seat belt, then mine, and we both headed inside. Jack and Mary were nowhere to be seen. We plodded through the corridor to the dining room. I could hear Ma and Pa quietly talking in the kitchen. It felt like walking through the house with an army, having Sean by my side. When we turned into the dining room, everything had been cleared away apart from mine and Sean's plates.

Sean patted me on the shoulder and went to sit. I tipped the quiche onto the empty spot on my plate, and cut it in half, giving him the bigger piece. Glancing at each other, we ate our Easter lunch, together.

• • •

THAT NIGHT, I lay on the sofa downstairs as everyone went to bed as if nothing had ever happened. Sean had checked on me before going up, but I said it was alright and told him he should get some sleep.

I did what I'd done at Christmas. I listened to the water running through the pipes, the taps squeaking, the floorboards creaking, the gradual onset of snoring. I felt the house go dark. I lay there until the early hours of the morning with just a single lamp on behind me. It must have been about three a.m. when I heard someone descending the stairs, and the sitting room door opening, slowly.

"What are you doing down here?"

It was her.

I had felt sick all afternoon, but as soon as I saw her, the nerves resurfaced with a vengeance.

Ma's head poked around the corner of the door. Then she moved into the room, into the dim light. She wore her dressing gown, and her eyes were swollen. Normally she wore her hair tied back, with bits sprouting out like wires into the thin air, but now it curled down over her shoulders in dark waves.

"You should be in bed," she said, shutting the door behind her, approaching me.

I sat up on the sofa. "I wasn't tired. I'll go up soon." I fiddled with the sleeves of my jumper.

Staring at me, she moved closer and narrowed her eyes. She had been drinking earlier in the day. I could smell wine on her breath, exuding from her pores, even the faintest whiff of it.

"I can't believe you've done this."

"What?" I crunched my legs into my body.

"You've turned Sean against me."

I waited in the air of her delusion. I felt nauseated as I always did near her. But I also, somehow, felt stronger. I didn't feel so afraid. I didn't need her. I didn't need this. I had another life, away from her, a life with Maya. And now I had Sean, too.

"Why do you hate me? Why me?"

As she looked at me then, I felt her jealousy; I felt her hatred, her resentment. I was the chosen one. I always had been. I wanted to know why, but I also was desperate not to find out.

"Stand up," she said sharply.

"What?" I said, hugging my legs.

"Stand up."

I got up off the sofa.

"Come, stand here." She pointed in front of her on the ground. "Stand here."

I stood in front of her. I didn't want to. I shouldn't have, but I did. It was like muscle memory, doing what she told me. My breathing got heavier. Coming through my nostrils like a horse going to battle. Hers did the same.

Rising.

Falling.

In.

Out.

"Do you want to know what I think of you?"

I shook my head, feeling terrified of the answer.

"No," I murmured, "I don't."

She sniggered.

"You and I," she whispered, her lips licking around the words, "we are just the same. I knew it from when you were wee. The apple never did fall far. And you, you'll also end up just like me, running away from things. Just you wait."

I shook my head at her again.

"I'm nothing like you," I said to her. "I won't ever be like you."

"You will be. You can never escape it. Just you wait."

My mother hated me for being the person she'd never been allowed to become. She punished me because she'd been punished herself, and she tried to ruin my future happiness just like hers had been ruined.

I couldn't let it come true. I wouldn't.

I felt her approaching, stalking me, I felt the air between our bodies dissolving.

And then,

using everything I had,

to break away from it,

I scrambled to the other side of the room.

I opened the door.

I ran up the stairs.

"Aisling. Come back here," I thought I heard her say. Maybe it was in my head. I couldn't be sure anymore.

I went into my bedroom and pushed my desk chair against the closed door. Then I got the suitcase out from underneath my bed and started to pack things into it. My clothes that I'd brought with me, the books that I wanted, one stuffed toy which I loved. I heard her knocking on the door, quietly, so as not to wake anyone else.

"Aisling," I heard her whisper, "Aisling, open the door."

I scanned the room. I felt my body trembling. I looked at the things I wanted to leave. Basically everything. The religious pictures, the hand-me-down clothes, the childhood memories. I didn't want any of that. I didn't want to remember this place. I wasn't coming back.

I heard Sean's voice outside.

"What's going on?" he said.

I did one last search around my room. Did I have everything I needed? Everything I wanted from this place?

I checked my bag: purse, passport, money, and Maya's handwritten address. The important things. I moved the chair, opened the door, and looked straight at Sean. I saw Ma's figure in front of me, quaking with both rage and fear. Sean looked at me like he'd just heard a terrible noise. His eyes sleepy, his top bare, and his checkered pajama bottoms draping down to the carpet.

"Can you take me to the airport?" I asked him.

27

ICICLE

Maya

London

ON EASTER MONDAY, I sat in the garden next to Mum. We perched on rusty cream chairs under the blossoming apple tree. Leaves and branches rustled, shuddering in the gentle breeze and weak sunlight. Mum was reading a book, and I was skimming through some of my revision notes. Steaming mugs of tea were balanced on the lopsided table in front of us.

All the mugs in our house were mismatched. Mum insisted on buying them from charity shops, and at one point we had about a hundred. Every time she saw a sweet one, she'd buy it.

"Well, it's for a good cause." She'd nod, tucking it into the stuffed cupboard.

My dad resented this obsession and had done a mass clear-out last summer, rolling his eyes and pointing out how ironic it was that he had to take them back to the charity shop. But lots of them remained. Many of them were patterned with little paintings of flowers or old villages or birds. The ones we were using as we sat in the garden were adorned with pink lacy patterns and yellow stripes.

Dad had just finished mowing the lawn for the first time that year, and the smell of the freshly cut grass hung around us. He was kneeling over the flower beds, weeding to make room for the bulbs to break through. In a month or so, the garden would be blooming with beautiful flowers, humming with bees, and dancing with butterflies.

On Saturday night, I'd seen Naomi and Tim. Ethan hadn't been there; it was still too soon. Even though I'd still been messaging Naomi and Tim regularly, it was much less than before Ethan and I had split. I'd been nervous to see them in person once again. We'd gone to a bar in Brixton. Chatting, drinking, smoking, I was surprised at how normal everything felt. As we stood there, I thought about how much I missed them, how much I missed the old times. But I had made the right choice; things were so good with Aisling. I just had to let time heal things.

Other than that, it had been a quiet weekend. It was just Mum, Dad, and me on Easter day. I had forgotten how at peace I felt at home. In the mornings, through the trees in the garden, the light streaked through the cracks like tunnels of sparkling dust. Laughter or the sound of the radio seemed to float through the house like a breeze in the evenings. I heard it in the shower, or when sitting in the garden on the spiky grass, or when pouring boiling water over a tea bag in the kitchen.

"What time is Aisling arriving again?" said Mum, blowing on her cup of tea.

"Her plane lands around noon. I'll leave soon to pick her up at the airport."

Mum nodded. I didn't look up from my revision notes.

"I think she'll be very tired, by the way. She said she only slept for a couple of hours in her brother's car before her flight," I muttered, glancing at Mum briefly and seeing her grimace.

"Do you know what happened exactly?" she asked.

I paused and looked towards Dad. He was using the small spade to get down to the roots of the weeds, extracting them. I watched the dead plants dangling there, covered in soil, twisting like broken ropes. He tossed them into a muddy heap.

"No," I said. "But I think it was bad."

Mum nodded again, her brow wrinkling into a frown as she guessed the subtext.

It had been the middle of the night when I'd spoken to Aisling. Amid the calm of being at home, I had felt worried about Ash. I was constantly thinking about her and what might be happening with her family, so throughout the day, and whenever I woke up in the middle of the night, I'd check my messages. That morning, I'd got up to go to the loo around 4:15 a.m., and when I'd checked my phone, I saw two missed calls from her. One at about 3:30 a.m., another around fifteen minutes later. I rang her back straightaway. As soon as she picked up, the tone of her voice told me everything I needed to know.

"Are you OK?"

"I'm fine. I'm OK."

"You're not hurt?" I was breathing heavily.

"No, I'm not," she said, pausing. "Maya, I'm sorry to ask, but can I come to you? My brother's just driving me to the airport. If not, it's fine, of course, I can go back to Edinburgh, but—"

"Please do. Please come here," I insisted.

"Will your parents be OK with it?" she'd said softly, her voice cracking.

"They will. I'll talk to them in the morning, OK?"

"Are you sure?"

"More than sure."

After we hung up, I didn't go back to sleep. I just lay there, feeling a mixture of relief, anger, and fear.

At the first sound of my parents waking that morning, I walked down to their bedroom in my pajamas and knocked on the door. They called me in. Lying in bed, propped up by pillows, curtains open, glasses on, newspapers and books spread out in front of them. I wanted to weep when I saw them, but I tried to stay composed as I explained things without including too many details. I asked them if Aisling could come here for a couple of days.

"Of course," Mum had reassured me, removing her reading glasses. "Is she alright?"

I'd nodded, swallowed. "I think so." Then I'd looked at Dad. "Dad? Is that OK? It'll be a few nights, that's all."

He didn't remove his glasses, so his big brown eyes were magnified. They looked wide and childlike, but eventually, he nodded.

"Thank you," I'd mumbled.

I thought about all this as I sat there in the garden. I kept replaying the whole thing in my head. I couldn't focus on revision notes. Mum was still holding her teacup in her hands, processing the information I'd given her as she peered up into the sky.

"I hope you don't mind her coming here?" I said.

"Not at all, she's welcome here."

We paused as we both looked at Dad, who was still weeding.

"You might need to be patient with your father, Maya. He's still adjusting, but he's coming around."

I gave a frail smile. I understood, and I was grateful for his efforts to understand me, too.

As I watched Dad, he sat back on his heels and wiped his brow with his muddy garden gloves. He was working hard to make a beautiful place for his family to enjoy, and slowly the weeds were fading away.

• • •

AT THE AIRPORT later that afternoon, I bustled through the crowds of people and looked for one face. Eventually, I saw Aisling emerge from the fuzzy streaks of humans who were passing me like flies. Through the distortions, her face came into focus; her body moved towards me, into my arms. I held her, listening to her heavy breathing.

"I've got you," I said. "I've got you."

Aisling and I caught the train and the tube back from the airport. We didn't talk much, but our bodies were always connected. She looked exhausted, her eyes purple, the usual brightness of them drained.

When we got home, Aisling put down her suitcase by the door and came to meet my parents in the kitchen. Mum asked how the journey had been, and Aisling politely replied that it had been no trouble. She asked how they were as she tucked her long hair behind her ears. I saw her earlobe that had been stitched all those months ago and my heart twisted. Mum responded that they were doing very well, then thanked her for asking.

"I'm grateful to you for having me, I'm sorry to interfere on your Easter weekend."

"Not at all, it's lovely to meet you," said Mum again. She flicked a quick glance to my dad, who still hadn't said anything. He was stood by the kitchen table, holding the newspaper under his arm, looking roughly in the direction of Aisling and me.

There was silence as the clock ticked. There was the clearing of a throat.

"Would you like a cup of tea?" my dad's voice croaked.

I looked at him, my body releasing slightly.

"Oh, um, I'd love a tea, thanks," she replied. I touched her arm reassuringly.

My dad nodded, popping the newspaper down. He filled up the kettle from the tap and put it back on the base, flicking it on. Everyone watched him.

"So, you're from County Clare, Maya tells us?" he asked, turning around, resting his hands on the back of a chair.

Aisling nodded. "I am."

"Some beautiful scenery in Clare, isn't there? Nice walks?"

Aisling took a deep breath and almost smiled. I had told her after Christmas that my dad hadn't quite come to terms with everything yet, and I was sure she was treading carefully around the dynamics.

"Very beautiful. Right in the middle of the Wild Atlantic Way, so some of the walks are spectacular, for sure," she agreed. "Like, the Cliffs of Moher or the Burren Way. Have you ever been?"

"I have, a long time ago now," he chuckled. "We walked the, um—" He frowned and put his fingers to his forehead. "Oh, what was it, the Cliffs of Kil—"

"Kilkee?" she confirmed.

"That's it."

Aisling nodded, and my dad smiled at her warmly.

"What tea can I get for you then, Aisling?"

"Anything caffeinated, thank you."

My dad almost laughed, and Mum and I flicked our eyes towards each other. I could never have imagined this happening. Having Aisling here, my parents meeting her. But all of it felt like a collection of small gestures which amalgamated to become more meaningful.

"Aisling, you can go upstairs and sleep if you'd like? I've just washed the sheets, they're all folded ready to go on Maya's bed," my mum said.

The implied permission for us to sleep in the same bed shocked Aisling. She stood there with her mouth hanging open, her purple-rimmed eyes fixed on Mum.

"Thank you."

Aisling took her tea, and we plodded up the stairs. When we got up to my room, I put Aisling's suitcase down at the foot of the bed. We gently unfolded the sheets as we put them on. They smelled of fresh cotton, the spring breeze, and apple blossom. We pulled them taut and smoothed them over with our hands, glancing at each other as we did.

She napped in my bed until dinnertime, curved like a fossil into one position in the crisp white sheets. I revised at my desk, looking around every now and again to see her mouth slightly open. I observed her heavy breathing and the fluttering of her eyelids. She slept through the smells of my parents' cooking, which floated through the house. Rosemary, onion, garlic. Before dinner, I had a shower and got changed, went downstairs to lay the table, and checked whether Mum needed any help. When I got back upstairs, she was still asleep. I sat near her feet and rubbed them gently until she stirred. As I watched her, my heart ached. I'd never felt this protective over another person before. I felt guilty for having this safe home, these parents. I wanted to give her everything I had.

"Dinner," I whispered. Aisling stretched her body.

"I could sleep forever," she said, yawning. "I feel so comfortable. Is this bed made from clouds or something?"

I smiled.

When we sat down for dinner, my parents put plates of warm, sweet-smelling food in front of us. Aisling smiled as she observed them. Her face looked pale and swollen from sleep. She asked my parents about what they did, about how they met. She listened to their responses intently, nodding, her blue eyes wide. In return they asked her all about her interest in poetry, her writing, and how she liked Edinburgh.

I felt so attached to her in that moment and quite in awe of her. After everything she'd been through in her life, even over the last twenty-four hours, she was so strong and kind and generous.

As we lay in bed that night, Aisling squeezed my hand tight, and I squeezed hers back.

"Thank you for having me here," she whispered.

"Thank you for coming."

I could hear her sniffing in the darkness, above the lull of the traffic outside and above the distant shouts of people on the streets. I clasped her hand even tighter. Something buzzed inside of me.

"I am here if you ever want to talk about—"

"Thanks," she interrupted.

I paused. "Really, anytime you want to—"

"Maya," she cut across me, her voice like an icicle dropping from a drainpipe, "you're not my therapist."

I felt myself, in the dark, release my grip slightly. I had an urge to get up and leave the room, to run down the stairs.

"Sorry," murmured Aisling, "I'm sorry. It's been a long day."

I told her it was nothing. We said good night and rolled over to face separate ways. I heard her breathing getting heavier as she fell asleep.

28

THE COLOR OF THE SKY

Aisling

London → Edinburgh

A FEW DAYS later, Maya and I traveled back to Edinburgh on the train. We played cards for a while, then sat reading our revision notes for a few hours.

I'd slept for about eleven hours the first night I'd arrived at Maya's. I felt such a sense of relief being there. It was as if I'd spent a couple of days being pampered at some fancy hotel, with nice-smelling sheets, scented soaps, and healthy food. I'd never been so looked after, not by anyone. And Maya, she'd had this her whole life. I found it almost baffling.

As much as I felt rested, deep down, I knew I was continuing to block things out. Not just stuff from the last few days, but from my whole life. I couldn't deal with thinking about anything in any detail. I was suppressing those thoughts and allowing myself to be elsewhere.

But the words my mother had said to me continued to swirl around my mind, like dust picking up into a billowing ball. I didn't want to run away from things like she said I would. I wanted to face things head-on, but what did that even mean?

When we arrived back in Edinburgh, it felt like we'd never left, but also like so much had changed. Waverley station seemed as grand and bustling as ever. The glass windows were cloudy and made the whole place unbearably bright. Out of the station, we caught a taxi back to Maya's flat. Maya chatted to the driver with that ease she always had with everyone.

She'd told me last term that she'd had terrible anxiety the past few years. She'd named it "the insect" and had even shown me where it normally lived in her body. I'd kissed her skin there, right in the center of her stomach.

Maya was one of those people who you wouldn't be able to tell felt anxious; she'd found ways to conceal it. She'd told me that for a long while drinking had been one of her ways of dealing with it. I didn't quite know how I felt about that, but when I thought about it a bit, I sort of understood. I guess I'd watched my mother doing that my whole life.

I stared out the window and looked lovingly at the streets of Edinburgh. There was something about this place. Something special. More than that, it was my home now. I didn't have another one to go to.

I'd need to get a job, I thought to myself. Ma and Pa had been sending me a bit of money to help me get by every month, but that would obviously stop. Plus, I'd spent a big chunk of my savings on the last-minute flight to London.

We got out the taxi, paid, and dragged our suitcases up the stairs inside the building. The big door to her flat opened and closed behind us.

"Gabe?" she called. We waited. Silence. Only the faint sound of the street outside.

"He must be out."

We turned to each other, thinking the same thought.

Afterwards, we lay in bed for a while, letting our bodies twist together.

"I might go for a quick run before dinner," I said. "I can grab some food on the way back?"

"Of course. I'll just shower and unpack."

I got changed, tucked keys and cards into my back pocket, and set off on my run, hurtling down the steps of her building and opening the door to the street. The crisp spring air of the early evening coated my body. I ran through the streets as the light began to fade into a pastel blue, grayness dimpling the sky like soot.

By the time I got to the meadows, the color of the sky was different from when I left. It was turning darker and moodier. My heart beat quickly as I ran to my spot and stopped, hands on my hips. I just watched and let my breath melt into a different, slower rhythm. I saw the towering hills around the city, the lines of trees, the walkways.

"Good to be back," I whispered to myself.

My throat went weak, and as the droplets of sweat from my forehead dissipated in the wind, droplets of tears came to my eyes instead. I wiped them away harshly, sniffing them in.

Then I kept running.

On the way back to Maya's flat, I stopped by the shop. She'd asked me to pick up ingredients for risotto. I walked around gathering them up, my pale skin shiny with sweat under the fluorescent bars of light.

The last thing I needed to get was wine.

Perusing the alcohol section, I didn't know which wine was right. I'd only bought one bottle of the stuff before, for Maya, that time when I went over for dinner. And even then, I'd had to ask the shop assistant which one was good. As I looked at the bottles, I wondered which one my mother would pick if she were looking.

Quickening my pace, I selected a random one and went to the checkout.

"Do you have ID?" said the lady at the till, scanning my items.

I rummaged in my back pocket, panicking for a second before handing her my driver's license. She glanced up at me, unimpressed. I was sweating more than I had done on my run. I felt uneasy and a bit jittery. Ma's face wouldn't leave my head. The lady scanned it through and handed me back my card. Without saying anything to her, I bagged up the items and left, walking swiftly back to Maya's.

"Alright!" I yelled when I got back into the flat.

I shut the door behind me. I could hear laughter from the kitchen. Gabe must be back.

"Ash, we're in here."

Rushing into the kitchen, I put down the bag on the counter and smiled broadly at Gabe. I was so glad to see him.

"Gabe! Good to see you!"

"Ash. Get in here!"

He hugged me.

"Oh God," I said, my head pressed up against his chest. "I'm real sweaty, sorry."

"I don't care. I just admire you for exercising; it's more than I've ever done."

Maya laughed, musically.

He cradled me in his arms. It felt deeply reassuring to be around him. I'd been intimidated by Gabe when I'd first met him, but over the last few months, I'd gotten to know him so much better from being at the flat so often. Gabe was one of those people who accepted everyone exactly as they were. I put my arms around him and patted him on the back. As I did, I thought about how Gabe would have been an amazing brother. I released from the hug, giving him a limp smile as I backed away.

Maya picked up the bag and looked through. She unpacked the shopping while Gabe asked me how the train journey had been. As I talked, in the corner of my eye, I could see Maya casting her eyes towards me after she removed the wine. She smiled faintly, sending me almost a nod of appreciation; she knew what might have been going through my head in that aisle of Tesco.

I went for a quick shower, just to wash off the sweat, and then we all stood in the kitchen as Maya cooked. We laughed and joked and caught up with Gabe. He told us all about his Easter weekend. Not that he celebrated Easter. Gabe was Jewish. He'd thought it had been a good opportunity for his mum to come up while Maya was away so they could be together for the end of Passover. She'd traveled from London to Edinburgh on the train and stayed in Maya's room for a few nights. He told us about how they went to Shabbat dinner at synagogue on the Friday, walked around town, and went out to a bar on the Sunday evening. As he spoke, I thought about the fact I couldn't fathom doing something like that with my mother.

I realized in that moment I would probably always feel like the odd one out. The only one who wasn't in contact with their family. The only one without anywhere to go during the holidays, without a childhood home to retreat to, without a family to celebrate birthdays with or to look after me if I got sick.

"Anyway," said Gabe, finishing his recount of the weekend, "it was great fun."

"And did your mum manage to meet . . ." Maya inquired nosily.

Gabe pushed his glasses up his nose and started to blush. "Yes, yes, she met Isla."

Maya squeaked excitedly.

"How was that?" I asked him, calmly.

"They got along well," sighed Gabe.

"Ash met my parents this weekend, too."

I nodded and pressed my lips together.

"Oh, yes!" said Gabe enthusiastically. "Isn't Amber just an older version of Maya? I adore her."

I stared at him, replaying his words in my head as he looked at me, waiting for my response.

"Oh. Absolutely," I whispered.

As I spoke, I realized I'd zoned out. Time had stopped. My mind had wondered backwards, and now the two of them were having a different conversation.

• • •

MAYA AND I didn't leave each other for another week, maybe longer. Time just seemed to pass without clear rhythm or pace. After we woke up, we would have coffee and breakfast, the sun piercing our sleepy eyes. We would sit there and yawn, kiss each other's groggy mouths, read the news. Given it was revision season, we spent hours each day at the library or in Maya's flat revising for our exams, which were coming up in May.

That week, I also applied for a job at the bookshop in Grassmarket. I sent an email explaining how much I admired the store, and that I'd love to pick up a few shifts a week as soon as possible.

A woman called Elsie responded to me the next day, saying that it would be great for me to join the team; they were actually looking for someone. But, she said, they could only offer me the job from about a month's time, when things started to get a bit busier. That was about after my exams finished, so I wrote back and accepted immediately. I could use the last bit of my savings until the job started.

One morning, the three of us were in the kitchen, making coffee in our pajamas.

"OK," I exclaimed, clapping my hands together, "I'm going back to my place for a few days today, I'm sorry I've stayed so long."

I'd thought about this. I didn't want them to feel I was overstaying my welcome, which I was sure I already had done.

Maya looked at me, tipping her head to the side. "Surely not."

Gabe turned around, too. "Aisling, not being funny, but why don't you just"—he looked over at Maya—"come and live here with us for the rest of the year?"

Maya widened her eyes at him. "Are you sure, Gabe? You wouldn't mind?" she said quickly, as if she'd had the same thought.

"No!" said Gabe, turning back to me. "It's great having you around. Plus, you'll have to move out of your accommodation over the summer anyway, but our lease runs until the end of September. But look, only if it works for the two of you, I don't wanna interfere."

As Gabe left the room with his coffee, Maya and I gazed at each other in disbelief.

"Well? What do you think?" She gaped at me.

I bit my bottom lip. "I mean, I want to, but what about rent? I can give a month's notice for my room at halls, and I only start that job next month, so it would be a month before I could pay you the—"

"Aisling, I'll cover it. Then, if it works, we can split it three ways next month or something. Look, as long as you're here, I don't care, we'll figure it out."

I stared at Maya's face. It made me feel safe, far away from everything else. I wanted to be here. With her, I was tucked away in a cushioned shell which clasped me in its curling walls, deep down on the floor of the ocean, the water rushing over my armor but not touching me.

That afternoon, I sent an email to the accommodation manager giving notice for my room. Maya put some music on. She boogied along whilst packing up her winter clothes into suitcases and putting them under her bed to make room for my clothes in her drawers. I unloaded the ones I had, folding them into the empty spaces of the cupboards.

"Hey," Maya said, shoveling one of the suitcases under the bed. "Look at this."

It was the life drawings.

"From our life drawing class."

I saw the sketch which I'd seen on her birthday, the one I thought was of me. Maya quickly tucked it away.

"What's that one?" I pointed at it.

Maya looked at the pile of drawings. After a moment, her shoulders sank.

"I drew this of you a while ago. I mean, from my imagination. God, that sounds so creepy! It was before we were—ah, God—"

She buried her face in her hands and I smiled, endeared. Then I laughed a bit, and she peered up at me, beginning to giggle, too.

"I saw it. On your birthday, when I brought you to bed," I admitted.

Maya gawked at me, her mouth dropping open, her eyes wide.

"I just saw it poking out from under the bed. I promise I wasn't snooping. It's very flattering!"

We both laughed and I kissed her on the head.

"I'll be back in a few hours."

Wheeling my empty suitcase out of her room, I headed to halls to collect the rest of my stuff.

• • •

AS I ENTERED my building, it felt cold and dark in the corridors, especially compared to the warmth of Maya's flat. I took my suitcase up the stairs. Before I unlocked the door to my room, I heard someone in the kitchen. I prayed it wasn't Toby starting to cook one of his meals, but then I heard a familiar voice singing along to something, presumably a tune playing on her headphones. I went down the corridor, suitcase in tow, and saw through the glass that it was Karen.

"Karen!" I said, opening the door and giving her a faint smile.

Karen jumped, slightly startled, before removing her headphones. She was sat at the table, eating her lunch of chicken salad and toast. Her green eyes widened as she registered me.

"Aisling, you're back! I haven't seen you in an *age!* How was home?"

"Yeah, it was good, thanks!" I lied. "How about you?"

I sat down opposite her, propping my suitcase up next to me. Karen nodded as she chewed on another big mouthful of food.

"Do you want some lunch? I have extra. Or you need to unpack?" She pointed to the suitcase.

"Oh, that's good of you, but I'm alright, cheers. I'm actually not unpacking. I'm gonna move into Maya's place for the summer, so I've come to get the rest of my stuff."

Those words felt so strange coming out of my mouth.

"I'll be sad to leave," I said, looking around the kitchen.

Karen frowned. "Me too," she said. "I'll miss you, Ash. We can still do coffees and stuff like that, though, right?"

I smiled at her. "I'd really like that. I'll be working at that bookshop down on Grassmarket from next month. Maybe we can grab a coffee around there sometime you're free? You've got my number. Come say hi anytime."

"I will." Karen smiled gently.

I went to my room, unlocking it and entering its hollowness. I stood there, assessing its contents. I felt sad, and a bit confused about why. Was I making a big mistake, moving in with Maya? I suppressed the foggy feeling of tears emerging. They were spurred on by the thoughts of me arriving here, by myself, sick with nerves, all those months ago.

I packed up the rest of my books, the candle, and my other bits of clothing, including the blue sweater I'd apparently stolen from Mary. Then I packed my bedding into my second suitcase, which I retrieved from under my bed.

My phone buzzed in my pocket. I got it out.

> Ash—let me know you got back to Edinburgh alright? And let me know how you're keeping. Sean

I stared at this for a long while, unable to respond. I appreciated Sean's concern but didn't want to be reminded of everything. Not right now. I tucked the phone away, back into my pocket. Before leaving, I looked at the empty shell of my room for one last time.

Back in Maya's flat, that night felt different. As we lay in bed, I kissed her lips, raising my body over hers and looking deep into her brown eyes. I stroked the top of her forehead with my thumb. Afterwards, lying next to each other in bed, fidgeting, neither of us could quite get comfortable. The sound of Edinburgh droned outside. Streetlights flickering, wind blowing, buses beeping, brakes squeaking.

Then, finally, we were still.

29

THE APPLE TREE

Maya

Edinburgh
MAY 2014

THE WEEKEND AFTER Aisling moved into the flat was beautiful weather. On the Saturday morning, we lay in bed and looked at the bright sunlight streaming in around the edges of the curtains.

"Shall we go for a walk? A break from revision?" I asked sleepily.

"Sure, I'd like that," murmured Ash.

We got out of bed, showered, then set off. Holding hands, we swung each other's arms around. Exams were coming up in a few days, but we couldn't have cared less.

Standing at the bottom of the hill, we looked up towards Arthur's Seat. As I balanced on a small rock, I kissed the top of Aisling's head. I smelled the scent of bedsheets still on her hair.

"It always makes me think of Mount Olympus, I dunno why," laughed Ash, propping her foot up on the rock and leaning her hand on her knee.

"Come on then," I said. "Let's go meet the gods."

Aisling's laughter fizzed in her chest.

We started walking up the hill. We'd packed sandwiches, apples, and two big bottles of water in a backpack, which she carried even though we'd argued over who would take it. The day was blustery, as it often was in Edinburgh. The wind was laced with the scents of crusty bread and wet grass, and the sun felt like it was bursting through the edges of the sky.

When we got near the top, we placed our hands on our hips and took in the view. The sweat grew into cold, thin droplets on our foreheads, and then the breeze whipped them away briskly. Our eyes squinted down to the land beneath. The small houses regimented and repetitive. The tall church spires and monuments dreaming upwards into the sky. The dark buildings. The patches of green in the middle of them. And in the distance, the lilting sea.

As we descended, we sat down on a rock, our hair whipping and blowing in our faces. We nibbled at our apples and tried to make out which buildings we could recognize, pointing and peering into the distance.

Ash looked at the rest of her apple. It was half bitten and browning.

"Did you know apple trees turn out to be different to their parents?" she said, looking at the bitten flesh and the seeds revealed in the core. I watched her. "They inherit different genes and characteristics from their parents so they can fight off different pests."

She took another bite and stared out, watching the clouds pass.

"If I throw the seeds here," she continued, "do you reckon they'll grow into an apple tree?"

"How long does it take to grow an apple tree?" I responded.

"Like, ten years or something?"

She picked the apple pips out of the core and placed them in the palm of her hand, assessing the weight of them.

"Then in ten years or so we can come back and see the tree?" she said.

I smiled at her as my hair whipped in front of my face. "I like the thought of that. We can come and eat the fruit," I fantasized, "like the Garden of Eden."

"Just without the guilt," she confirmed.

I knew what was happening, at least I thought I did. I hoped. I was sure this was her way of beginning to break free of everything.

Looking at me straight in the eye, her mouth curled up into a smirk. Hair whipping around her face, she knelt, found a spot in the earth, and dug with her fingers, manically almost, tucking the apple seeds into the ground and patting the soil back on top.

I watched her desperation. It reminded me of someone who was digging up the skeleton of a long-lost lover or, perhaps, a long-lost family member. Searching for them, wanting to face them, to say goodbye, but simultaneously letting go, piling the soil back on top, like wet ash after a fire. Burying them. I felt like we were, in fact, at a graveside. This digging and burying would always haunt us.

Aisling stood up and looked at where she'd planted the seeds, clapping her hands together to remove the dirt.

"We have to remember where it is." She studied our surroundings. "Do you think you can?"

I nodded my head. "I'll remember."

• • •

AS WE WALKED back down the hill, I wanted to ask Aisling about what had just happened, about what she was thinking. But

I knew I had to wait until she was ready. I'd felt her snapping at me that night in bed, at my house in London, when I'd asked her about what'd happened over the Easter weekend. I didn't want to rush her again. I felt the insect stirring inside my arms and throat, but I stayed quiet as she walked ahead of me, traipsing down the muddy and mossy slopes, over the bumps and around the dips.

Halfway down, she stopped and turned. I stood above her on the slope of the hill.

"I'm sorry," she said. "I just . . ."

Her breathing was heavy. Her eyes full of quivering blobs.

Her body broke down and she wept. She knelt to the ground. I knelt with her. I wrapped my arms around her as she shook, the small ball of her being quaking. I wiped the tops of her cheeks, taking the tears into my own hands. As time dictated it, we moved from kneeling to sitting, our legs tucked into our chests. I embraced her, and her head nestled in the crook of my neck.

We sat there silently.

Would she let me see her darkest thoughts, and would I be able to bring some light?

I shut my eyes and thought of a dark cloud rolling over, the rain, and the arrival of the sun. The creation of a rainbow.

"Do you think I'm going to turn into her? Do you think she was right? Am I going to be just like her?"

I paused. This was the first time Aisling had offered a window into what her mother had said to her.

"Aisling, I think you're like the apple tree."

Hearts beating. Separately. Together.

"Your mum projected everything onto you. Those are parts of her, they're not parts of you."

I spoke loud and fast, above the rush of the wind. I hoped that I was right.

Ash looked at me, her eyes full.

"You're the apple tree," I repeated.

"I'm the apple tree," she whispered back to me, her voice echoing in the wind.

30

UNDER THE LIGHTS

Aisling

AT THE END of May, exams were finally finished. They'd been difficult to get through, especially as I'd been so distracted that I hadn't managed to do much productive revision, but at least they were over.

Maya and I had continued to go to poetry society throughout the exam period. It was good to see everyone and take a break from work. I felt like I'd become less reserved because of poetry society. I didn't know these people all that well, but they had grown familiar to me. I cared for them. I saw them every week at meetings and at socials. And there's something about having people read or listen to your writing that makes you feel close to them. They know about intimate parts of you, sometimes the deepest, darkest, scariest parts.

I hadn't done much writing since before Easter. The stuff with my family had made me feel scared about what might come out if I tried to write anything, and I wasn't ready for that.

When we'd been for that walk up to Arthur's Seat a few

weeks before, I'd broken down in front of Maya. I felt so embarrassed after. She'd seen how scared I was, but I had exposed just a small part of my fear to her and I knew there was so much more inside of me. I didn't think it was fair on her to make her deal with all that, though, so I constantly focused on tucking it away. But it was coming out in different ways; I was becoming increasingly frustrated over the smallest of things, and sometimes I'd take it out on her, snapping at her or just being grumpy. I felt bad about it. I tried to control it, and I'd always apologize after, but the fact it was happening more scared me.

I was settled into Maya and Gabe's place properly by that point. It was great living there—seeing Gabe and Isla often and getting to be with Maya all the time. I should have been perfectly happy.

Sean had been messaging me a lot as well, asking me if I was OK, checking on me. He'd even tried to call me multiple times and I hadn't picked up. It had been weeks of him trying to contact me and me not responding. I could tell he was getting worried. His last one read:

> Please just let me know you're ok Aisling—
> that's all I want. Don't take this out on me.
> I just want to protect you.

I couldn't deal with responding. I just couldn't bring myself to type something back to him. I appreciated him reaching out, but I didn't want to be thinking of anything to do with home, so every time he messaged, I'd read it, then tuck my phone back into my pocket.

A couple of days after my last exam, I had my first shift at the bookshop.

I had a shower in the morning, got dressed, and kissed Maya

on the forehead before leaving to walk to the shop. She'd finished exams the day before me but was still sleeping it off.

It was a Saturday. A bright, delicate late-spring morning. As I walked over the deserted meadows, birds spun in the air like acrobats. Their wings flipped and swerved through the trees. The fresh grass fluttered in the breeze, and the leaves above looked so full and healthy. The blossom trees bloomed as well, their fragrant fuchsia petals hanging there delicately. Everything seemed refreshed and born again.

When I arrived, Elsie, the shop manager, greeted me with smiles and rosy cheeks. A long green shawl draped over her shoulders, and gold sparkles caked her eyelids. Although we hadn't talked much, I recognized her from when Maya and I had come in before. As she started to take me around the sections of the shop, I listened quietly, even though I knew every single nook and cranny of the place. We went downstairs to the poetry section and came back up, finishing off with the modern literature.

"So, that's all there is to know," she said, catching her breath from walking the stairs. "You're studying English and Scottish literature?" She smiled widely, her hands on her hips.

"Just finished my first year."

I paused.

"I have to say, I come in this shop a lot. I really love it here."

Elsie squinted at me. "Yes. I thought I recognized you. Well, that's good. We like it in here, too. Now," she said, "let me show you how the till works."

• • •

IN CONTRAST TO my exams, which had made hours feel like weeks, my first day working at the bookshop passed quickly, like a flicker of light. I got to sit and read at the back of the shop when no customers were around, or just sit and admire the place: the spines of the books on shelves, multicolored ridges of bound-up knowledge; the worn-away carpets and creaking floors; the higgledy-piggledy-ness of it all.

When people purchased books, we'd chat a bit while I put them through the till. Sometimes people had questions. I should have had more trouble, with it being just my first day, but the truth was, I knew the shop so well that it was always easy to answer them all. I could see ghosts of Maya and me from the past in that place. Something inside me missed those times. Weirdly, I felt myself getting further away from her even though I was with her almost all the time. But then, I told myself, I felt far away from everyone at this point.

After I finished my shift, Elsie taught me how to close up. She'd been supervising me all day and taking care of the new stock.

"You did well." She beamed.

I smiled and told her I'd enjoyed myself.

When I finished and sauntered out onto the street, the light outside was soft and golden. Edinburgh Castle was lit up by honey-colored rays of light. I checked my phone to see a message from Maya.

> Ash—we're in that bar on Cowgate, come
> join when you're done with work! X

There was the final poetry society social of term that night. Meetings were stopping over the summer, even though most people were sticking around Edinburgh over the coming months to work, including Maya and me.

Cowgate wasn't far from the shop at all. I'd just drop in and say hi to everyone, I decided. It'd be nice to catch up with people.

I walked down the street and under the bridge to enter Cowgate, a damp, dark street which was cave-like. Dripping and mossy.

I crossed the road and entered the bar. It felt like a cinema. Its alluring darkness, its deep red chairs, its wooden booths. The floor was like lava, splashing light up onto the walls towards the ceiling and the bottles behind the bar. Bright green and pineapple-yellow beams projected rigid, jagged shapes, but the rest of the place was dark. It stank of sharp alcohol. The music was some sort of mindless drivel, and it was loud enough that you couldn't hear yourself think.

I'd never felt at home in these places, but Maya always had.

I spotted them all and walked over. Maya was wearing a sparkly top, bright blue eyeshadow, and big hoop earrings. I was jealous. I couldn't wear earrings anymore since Ma had split my earlobe and wearing only one made me look sort of like a pirate. Her hair was bunched on top of her head into a bun which let her curls tumble down. I could see Harry, Isla, Isabel, Tommy, and Rohan were there, too. They were all laughing, joking around with one another. As I approached, Maya put her arms out.

"Ash! You're here!" she yelled.

She stood up and kissed me on the lips, holding onto my neck. I could feel her fingernails digging into my skin. I could smell the vodka and bitter lemon on her breath. I pulled away, slightly embarrassed, my stomach twisting.

"Alright, everyone," I said, poking my head around Maya and raising my hand towards the others. They all smiled and greeted me.

"Come on," said Maya, yanking my hand. "Let's get a drink."

She dragged me towards the bar, and when we got there, she spun me around.

"What do you want?" she slurred. "White wine?" Her eyes darted all over the place.

"Maya, are you alright?"

She looked at me droopily. "Yeah! I just wanna get you a drink."

She leaned up on the bar, trying to get the attention of a bartender. I stared at her, thinking about how I hadn't seen her this drunk in ages, maybe even since we'd started going out. I wondered why she was like this now.

"I don't drink, Maya, remember? I'll just grab a water or something."

"Come on, Ash, surely just try it."

I stared at her, confused. "Nah, I don't fancy it."

I turned towards the table, to check no one could overhear. They all knew I didn't drink, sure, but I didn't want to draw their attention to it even more.

"It's just a fucking drink, Ash." Maya waved at the guy behind the bar.

"What can I get you?" he said, throwing a tea towel over his shoulder.

"A water would be great, thanks," I said quietly.

"No, no, no," Maya insisted, pushing me away, her skin on my skin. "Get her a vodka lemonade."

I kept quiet while he made me the concoction. I could feel my frustration bubbling, about to explode. It wasn't just frustration; it was also confusion about what had brought this on. What was she doing?

"I'm buying," Maya said to me.

"You're too drunk, Maya."

She shook her head, closed her eyes. "No, no," she slurred.

A flash of light over her face. Over her eyes. A single strip of light.

The bartender put the drink on the counter, and Maya paid. He nodded his thanks, and she handed me the drink. After a beat, I took it.

"I just want you to have fun," she mumbled, touching my face.

I moved her hand off my cheek. "I can have fun without."

But without listening, she grabbed my hand and tugged me over to the table, dancing to the music in the background.

In the interest of not making a scene, I sat down next to Tommy and started to sip at the vodka and lemonade. I'd never tasted vodka before. In fact, I'd never tasted alcohol before. It burnt my throat and I didn't enjoy it, really; it wasn't nice. Maybe that was why my mother liked it? Was it a sort of self-imposed punishment?

I chatted to Tommy, trying to ignore everything else that was going on around me, trying to suppress my building irritation. Tommy was the same year as me, also doing English. He'd just finished exams as well. We spoke about how they'd gone, and I asked what his plans were for the summer. I watched Maya across the table. She was chatting to Harry and Isla. She seemed different, almost unrecognizable.

After a while of occasionally sipping at the horrible drink, my head started to feel like a balloon. I tried not to dwell on the sensation too much, but the feeling was disconcerting. Tommy was telling me about how he was planning to work at the Edinburgh Fringe this summer to make some money, but I couldn't really concentrate on his words. I just kept feeling myself get all floaty, and glancing over at Maya, watching her also transmogrify in front of my eyes.

"Oh my God!" shouted Maya. "I love this song!"

Standing up from the booth, she seized my hand. She towed me into the corner, where there was just enough space to dance. Maya held my hand above her head and started to wiggle her body. Her eyes sagged, and her drink spilled over the already sticky floor and over her top. The lights flashed over her like a flicker of moonlight from a crack between a pair of curtains. Her eyes watched me. Everyone at the table giggled.

"Come on, guys," shouted Maya. "Get up and dance!"

I felt dizzy and a bit sick. "I'm gonna go home."

"Stay," she begged me, drawing me in towards her, her eyes lolloping.

"Why are you doing this?"

"Come on, Ash," she said. "It's just being young, forgetting about everything."

It was funny how her technique for forgetting was mine for remembering.

As we stood there, shuffling our bodies around, people jumping near us, I watched Maya.

I couldn't.

"I'll see you at home," I whispered in her ear. I don't think she even heard me. I walked away from the group, stopping by the door to take one last look at her sipping her drink as she spun around under the lights.

31

A STORM

Aisling

JUNE 2014

A FEW DAYS later, I sat behind the till in the bookshop, working a shift. It was raining hard outside. The kind of downpour that would soak you through to your bones in an instant and make you shiver until you'd had a warm bath and sat in front of a blazing hot fire.

After that night at the bar on Cowgate, Maya had come back in the early hours of the morning. I'd heard her bash her way into the flat, go to the toilet, and clamber into our room. She'd got into bed, fully clothed, and snuggled up right next to me, curving her arms around me. I'd been wide awake, but I pretended to be asleep. I could smell cigarette smoke, sweat, alcohol. It didn't take long for Maya to drop off. I could always tell when she was sleeping because her breathing got loud. On the verge of a snore, but not quite.

In the morning, I got up, went for a run, had a shower, and when I returned to the bedroom, she was sat in her desk chair wearing a big jumper, her legs tucked up into her body. She looked at me like a scared child. I stood there, in my towel, with my soaking-wet hair dripping on the floor.

"I'm so sorry. I don't know what got into me."

"What do you mean?" I responded, wondering if something else had happened.

"I should never have pressured you like that, and I shouldn't have got that drunk. I just, I've been—" She paused. "I'm sorry."

I could feel her guilt. I could feel her disappointment with herself. I waited for a moment, looking at her. Looking at her familiar face, her beautiful eyes.

"It won't happen again," she assured me.

I nodded and smiled, avoiding asking any of the questions I wasn't certain I wanted the answers to. Instead, she spoke again.

"I want to support you. I promise I will. I want you to tell me how I can—"

I went over to her and kissed her, and she peeled off her jumper as I dropped my towel.

As I sat there in the bookshop, I watched the water stream down the glass of the shop window and read a collection of poems by Sylvia Plath. I could feel the chills of wind blasting in through the door every time it opened. The cold seeping into my skin and through to my joints.

Given the terrible weather, the bookshop was extremely quiet. Every now and again someone would come in, absolutely soaked through. They would put down their umbrella and drip all around the shop, clearly entering for the sole purpose of shelter. They'd pick up some random book, pretend to be interested, and then leave without making eye contact with me.

As I read Plath's poem "Mirror," I heard the bell on the shop door tinkle as it opened. A bluster of air on my legs, the patter of the rain outside louder for a second.

I looked up, expecting another shelter-seeker.

A lad with no umbrella, wearing a navy blue waterproof coat, big builder's boots, and jeans. Sopping wet. He drew back his hood and shook out his hair in his hands.

It couldn't be?

But it was.

It was Sean.

I blinked, hard.

"You're alive then." Sean smiled.

I shut my book. "What're you, how did you . . ."

He shook out his hair again with his hands.

"Had the address of your student halls from ages ago," he said, wiping the rain off his face. "Went there, waited outside like a complete fool. Asked random people going in if they knew you." He sniffed and shrugged. "Eventually this girl with mousy blond hair, Carol?"

"Karen," I corrected him.

"Yeah, alright, Karen or something. She knew you. I said I was your brother. She looked a bit worried. Didn't know your new address but that you might be working a shift here, so I thought—" He let out a sigh. "I thought I'd give it a try."

• • •

SOON AFTER SEAN arrived, I went to find Elsie and asked if I could take my lunch break a bit early today.

She nodded. "Everything OK?"

"My brother's just here from Ireland," I said. "I didn't realize he was coming, so . . ."

Elsie looked at me with kind eyes, realizing my fluster.

"Go," she said. "Take the afternoon off, hen, I'll cover it."

"Thanks, Elsie. Honestly, thanks."

Sean and I sat in the coffee shop down the street. The windowpanes were gray with condensation, and drips carved their way down to reveal more dribbles of rain on the other side of the glass. People outside in the wet bustled their way through the streets under umbrellas or newspapers, homeless people sheltered in doorways, and the market sellers huddled under their stalls. Inside the café, the steamer blasted and cooed, and the milk frother buzzed in the background.

It wasn't busy, not on a day like this. There were only a few others in there, sat at the wooden tables. Sean had ordered us two cappuccinos.

"Here you go," said the man, popping down the turquoise mugs in front of us.

"Thanks," said Sean, his hair still wet. He took a sip. I didn't touch my coffee yet.

"I flew out to see you. I've got a return for this evening. It was just, you know, with you not responding to my calls and messages and all, I thought I'd just check you were—"

"I'm sorry," I muttered. "I should've got back to you."

It must have cost him a bunch to fly out. I felt guilty. I could have avoided all this by just messaging him back or picking up a call. I sat there, my hands in my lap. I didn't know where to start.

Sean pressed his lips together and wriggled around in his seat. "Don't be sorry."

"I couldn't handle it all. I've been trying to process it all and I . . ."

He nodded.

"Aisling," he said after a long silence, "it was my decision to come out here. I was just worried about your mental . . . state, after all that. I didn't know what sort of mindset you'd been in, and I'm sorry for just showing up; I didn't know if you were gonna— I was worried, that's all."

"I'm sorry, I didn't mean to make you worry."

Sean cleared his throat. "But you're not in that sorta place, are you?"

I shook my head, sensing his implication. "No, no, I'm not. But I have been . . . struggling."

That was the first time I'd said it out loud, to anyone.

He took another sip of his coffee. "Have you thought about going and seeing someone?"

I stared at him, blankly. "What, like a therapist?"

"Yeah, like a therapist. To talk. I could help you with cash if you need."

I took a sip from my drink then, and felt the tepid liquid run down my throat.

"Sean, I can't believe you'd—" I shook my head. "Thank you, but I think I'm OK."

I knew I was struggling, but I'd never known how to ask for help. I'd been commanded not to tell. Don't tell, that's what she'd instructed me. Talking to people always scared me. That was why speaking to Maya, even all those months ago, had been such a big deal for me. People always ran away, or let you down, or might not be able to help. I'd never even considered going to a therapist.

Sean gave me an encouraging smile, pressing his lips together again.

"Just don't bottle things up, alright? And let me know if you need help. Text me back, for God's sake. I am here, despite my cowardice the last years."

We waited for a while and let the clanging of crockery and the faint music in the background hang in the air around us. Sean took a breath.

"I hope you forgive me. I was an idiot. But you don't need to associate me with them anymore. I really wanted to say that to you and to know you'd heard me. I won't be seeing them."

"Thanks, Sean."

We waited, nodding at each other.

"I just don't really understand," I said, finally releasing some of the tension from my body. "Why me?"

Sean looked at me, his eyes growing bloodshot.

"I don't know," he whispered. "Perhaps she felt you were . . ."

He stopped himself from talking and stared at me with wide eyes. I let his unsaid words run over me like a waterfall of soil.

I knew what he was thinking, what we were both thinking. I didn't know how he was aware of Mother's sexuality, but in that moment I knew that he was. Pa knew, too, I was sure of it. I was certain that was where his blindness towards everything had begun. I knew about it because she'd confessed it to me, months ago, as she lay there on the sofa after spying on Orla and me through the crack in the curtains. I could see it in her eyes then.

But really, when I thought about it, I'd known long before that confession. I'd always suspected that Mother had been forced to suppress who she really was by her own parents. Her relationship with them, before they died, had been hostile and toxic. Even I could see that being around all of them together when I was wee. I remembered clearly that she had not been sad when they died. There might have even been a glint of happiness in her eyes.

Perhaps they'd found out somehow. Perhaps they'd done awful things to her. Perhaps they'd treated her the way she treated me. Even though I'd never known the details, I knew it had something to do with them, and I had long wondered if she was projecting their disapproval onto me.

I shuddered and broke eye contact with Sean.

Then I cleared my throat, not wanting to go further down that path.

"So, what, you're just not to see them anymore?" I asked.

"Nah, I don't think so. I don't think Mary and Jack are there yet, but I'm hoping they'll come around soon enough. I'm trying to start over a bit. Get some distance from them to focus on other things."

Sean smiled in solidarity.

"I'm actually engaged," he said.

My mouth dropped wide open; my hands flung out from under the table and reached over towards him.

"Shit!" I said. "Sean, that's great news! Who to? How long? When's the wedding?"

"She's called Bridget. I met her in Dublin. Works at UCD in the School of History. I mean, you'll definitely get along, I just know it," he said, smiling. "I think it's made me see things differently, gradually, over time, being with Bridget, talking to her about our family, about you. It let me become someone away from home, and it allowed me to get some perspective. Anyway, I was wondering if you and . . . Maya, is it? I was wondering if you would like to come to the wedding? It'll be next year sometime, probably in the summer. Very small, and I won't be telling the parents, don't worry."

I grinned and tipped my head sideways, putting my hand on my heart, which ached with happiness.

"I wouldn't miss it."

Sean's eyes were wet, and he sniffed. We sat, staring at each other, and he reached across the table, patting my hand with his.

"And what about Maya. Can I meet her today?"

"Oh! I think she's busy," I lied, pulling my hands away slowly to pick up my coffee cup. "And if your flight back is tonight, then I think you'll miss her."

"That's a shame." He frowned. "Well, another time."

As I swallowed some coffee, I realized I didn't want them to meet. My world of home and my world with Maya. They were colliding already, and I didn't like it. It was creeping up on me too quickly, like a lion stalking its prey. A storm was brewing. It had flashed the other night like lightning.

I'd seen her.

She was coming for me.

She was coming for my safe space, my protection.

Ma.

Maya.

I could feel it, and I'd do everything I could to stop it.

32

I FEEL

Maya

London
JULY 2014

MUM OPENED THE door to the house as Aisling and I stumbled up the pathway, lugging our bulky bags on our shoulders. The light of the sky was like a ripe plum. It was one of those summer evenings where even after the sun has gone down, it pokes its head around the horizon, creating a bruised evening light.

Mum beamed at the two of us. She wore her overalls and a thin cardigan, her glasses balanced at the end of her nose. As I walked towards her, she opened her arms wide.

"Welcome home." She hugged me, kissing the side of my head.

She turned to Aisling as well, and awkwardly, Ash placed her bag down outside the house and embraced my mum, her body rigid.

"Good to see you, Amber, thanks for having me again."

"Not at all, great to have you. Supper will be ready soon."

We followed her through to the kitchen, where Dad was cooking.

"Ah, girls, I'm glad you're here, food's almost done."

Aisling and I had come down to London for a long weekend.

The summer was racing by, and we hadn't seen my parents since Easter. In August, I was going be working at the Fringe in Edinburgh, so we wouldn't have the chance to visit when that came around.

Since exams had finished, Ash had been working long shifts at the bookshop, and I had been doing some writing, drawing, and reading for next year's classes. Between our schedules, Aisling and I had been seeing less and less of each other, so I thought this trip would be a nice chance for us to spend some time together.

"How was the journey?" asked Dad.

"No trouble at all." Aisling nodded.

Mum grabbed a bottle of crisp white wine from the fridge. "Wine with dinner?" she asked.

"Oh." Aisling jumped. "I'm alright, thanks."

"OK," said Mum, getting three wineglasses from the cupboard. "Well, I can offer you some orange juice or apple juice?"

Aisling paused. "Orange, please."

Mum nodded, retrieving that from the fridge, too.

Ever since that night at the bar in Cowgate in late May, Aisling had become even more uncomfortable whenever I drank alcohol. I had been an idiot that night. I felt so bad about it. I wanted to tell her that I'd only gotten that drunk because I was nervous. I felt things changing. I had sensed her becoming frustrated with me, and that had made me anxious and scared. I only wanted to escape.

But I hadn't told her all that. I had simply apologized. She'd been going through so much, and I didn't want to bother her with my own feelings. Ultimately, all I wanted was for her to be happy.

Since that night, though, I could tell that Aisling saw me differently. It wasn't just when I drank alcohol; it was when I did other things, too. Sometimes it was as if she didn't recognize me at all anymore. More than this, she'd started snapping at me in those moments. One day earlier in July, we'd been in the kitchen together after dinner. I'd put on the washing-up gloves and started to scrub at the dirty pots left in the sink. Aisling watched me as I washed the big saucepan in hot water, cleaning it rapidly, the steam weaving its way up into the air around us. And when I turned around, she was gone. I went to the bedroom and found her, half laughing as I told her I'd been talking to her and only realized she wasn't there when I'd turned around.

"Sorry," she'd said awkwardly.

"What's going on? Did I say something to annoy you or—"

"No!" she'd hissed, her voice cutting me with its sharp edges. "You didn't do anything."

The insect had engulfed all my muscles. It wrung them out, sapping the life out of them. Recently, it had come back bigger, stronger, and with more stamina than ever.

Aisling was pushing me away. She hadn't even told me that her brother had come to visit her; I'd found out from Gabe.

"That's a bit of a slap in the face to find out from Gabe instead of you, Ash," I'd said to her at home one night in our room after dinner. Gabe had just gone out.

She sat on the edge of the bed, peeling her tights off. "I didn't think you'd care."

"Of course I care. You don't want me to be involved in that part of your life, do you?"

"Maya, he was only here for a few hours. I didn't tell you because I forgot, I'm sorry, OK, can we leave it?"

After taking some time away from her that night, just sitting in the other room, I'd left it. I knew she hadn't forgotten; I knew that was a lie, but I didn't want to get into an argument.

Aisling was becoming ever more distant from me, ever more reserved about her feelings to the point that I'd given up asking. But I was doing the same; I wasn't letting her in or telling her about how I felt. We were closing off from each other.

But the thing was, she still had that effect on me that she always had done. Sometimes she still made me feel like I had been lifted from this planet and was gliding up towards the stars. But it was the joy, our joy, that was fading.

I knew this would pass; it would just take time. She'd been through trauma, and I didn't exactly have my shit together either. Even if sometimes I did want to leave the flat, slam the door behind me, and run away, I knew, deep down, that we'd be OK. I just didn't want her to slip further and further away from me, or from that sparkle of life which I had so often seen in her eyes before.

"Oh, by the way, Aisling," my dad said excitedly as he started to serve up the supper, "before I forget, I came across this book that I thought you'd enjoy. It's about Irish female poets from the eighteenth to twentieth century. I got it for you. It's upstairs, I'll go grab it after dinner."

Aisling swallowed a sip of orange juice, almost choking. She peered at him with wide eyes, not knowing what to say. I knew what she was thinking.

"That's so good of you, Richard. Thank you."

My dad finished serving up the food, and Aisling began to carry the bowls through to the dining room. I felt my phone buzz in my pocket. Quickly, I glanced down at it. My face went hot as I saw it was a message from Ethan.

> Hey Maya. Hope you're well—I know it's
> been a while. Naomi told me you're in

London for the weekend. Did you want to grab a quick coffee, maybe tomorrow? It would be nice to see you. X

I stood there, reading it, rereading it, blinking to check if I was seeing things properly.

"You OK?" asked Aisling, touching my arm. "Are you not coming through?"

I looked up. The food had been taken into the next room. My parents were in there, chatting and laughing; I could hear them. It was just the two of us in the kitchen.

"Yeah." I shook my head. "Sorry, let's go eat."

• • •

THE NEXT MORNING, I waited in a coffee shop for Ethan to arrive. In front of me I had a cinnamon bun and two coffees, but I wasn't hungry or thirsty. I was reading a collection of Mary Oliver's poems, but I couldn't focus on it; I felt light-headed and sleep-deprived.

Once we had got upstairs and slipped into bed the night before, I'd told Aisling about the message from Ethan. She'd asked me why I hadn't told her sooner, and I'd responded that I'd only just got it.

"Do you not want me to go see him then?" I'd whispered to her.

"No," she said, "it's fine, go. I don't mind, I can just stay home and read or whatever."

I waited in the silence between our bodies.

"Thanks. It's just that obviously I've been waiting for him to reach out when he's ready to be friends, and before we went out,

I mean, he was one of my best mates. I've known him since I was little, and it would be nice to—"

She cut me off. "Maya, it's alright, go see him."

"Thank you."

"It's OK."

I'd left the remnants of our words hanging in the air for a minute, then kissed her on the lips, softly, and rolled over. I barely slept that night.

As I waited there in the café, thinking about Aisling being sat at home reading her book, Ethan walked in. I saw him and folded down the page on the poem "The Whistler." He wore a loose shirt, linen trousers, and white trainers. He smiled at me, dimples in his cheeks, and gave me a short wave. My heart shifted, seeing him.

I got up from my seat. We hugged, holding on to each other for a bit longer than was normal. We finally released each other and sat down, both of us laughing nervously. He made himself comfortable in the chair and then looked at me, his mouth hanging open in a smile.

"Thanks for the coffee."

"Don't mention it," I said.

"So," he sighed, pausing as we laughed a little again, "how've you been?"

I nodded but didn't speak for a second. I felt like my throat might crack, and I didn't know how to summarize the last seven months.

"I've missed you."

The words came out of my mouth before I could even think about them.

His lips pressed together. "It's good to see you, Maya. Really good."

We sat there and chatted, easing back into each other's presence. I asked him how the rest of his year at Cambridge had been, about how his parents were, about how his summer was going. He reciprocated, asking me questions on similar topics. I tried to block out the fact that a year ago, things had only been just beginning with us. The kiss in his kitchen, Lucy's birthday party, the picnic on Hampstead Heath. But it all felt so distant. I was a completely different person now.

"Do you have someone at the moment then?" Ethan asked after a pause in the conversation. He played with the handle of his coffee mug.

I cleared my throat. "Yeah. I mean, yeah, I do. How about you?"

Ethan took a big breath in and out, then slowly shook his head. "Not really. I mean, there's been a few things here and there, but nothing serious, no."

I looked down at the table and ran my fingertips over the surface.

"Do you think we might be able to be friends?" he asked me, narrowing his eyes. "I think I'm ready. Just about," he laughed. "I mean, it's weird seeing you, but I'd like it if we could be mates."

"I'd like that, too," I muttered.

He swallowed, then took a long breath of relief, leaning in towards the table, clasping his hands on the surface. I patted his knuckles. Ethan brought his other hand over mine, and we sat there, our skin connected.

• • •

WHEN I GOT back to my parents' house about an hour later, I took the stairs two by two. I felt so strange and a bit confused. I felt happy, but guilty about it, and most of all, I felt anxious.

Seeing Ethan had been great; I was so pleased that we could try to be friends again, but it had only brought home to me how distant I was truly feeling from Aisling.

I burst into my bedroom to see her lying on my bed, reading the book my dad had bought for her.

She smiled and sat up, saying hello, her face melting like a glacier thawing in the sun. I hastened over to her as she lay there, and I kissed her. I let our lips meet softly; then I pressed harder, stroking her neck and hair, her arm. I pulled back and looked into her deep blue eyes, breathing out and sitting down next to her on the bed.

"What's wrong?" she asked timidly.

"I'm sorry," I said, rubbing her bare arm with the palm of my hand, gently. "I'm sorry things have been strange recently, I'm sorry I left you here today, I'm sorry if I've been annoying you recently, I'm sorry if the drinking bothers you. I can stop or cut down or whatever."

Aisling stared at me, asking me with her eyes what had come over me. Her nostrils flared.

"What did Ethan say?"

I shuffled as I sat there. "Nothing really, he just asked if we could be friends, I told him about you, we chatted about life in general."

Aisling nodded and shut the book, putting it next to her on the bed. She rested her hand on the cover and tapped her fingers on it.

"You don't need to apologize," her voice murmured.

I looked between her pupils. It was as if she hadn't heard what I'd said, as if she hadn't understood what I was trying to do.

"Aisling, I—"

"It's all OK, Maya."

After a moment of looking at her, I spoke again. "I want you to talk to me. I need you to tell me what you're feeling. This won't work unless . . ."

As our eyes connected, her nostrils flared again. Sometimes I felt like I didn't recognize her. It was almost as if she were becoming a different person. A person she feared, a person I feared.

"I don't even know, Maya."

"Try," I said, cupping her face in my hands. "Please, Ash. Try. I want to make this better."

"I feel . . . I feel that I love you," she said, licking the wetness which had started to run from her nose. "I feel far away from you, from the world, from the present. I feel scared, angry, guilty, jealous, engulfed in darkness. I feel awful that I've been so short with you recently. I feel sorry . . . I feel . . ." Aisling's eyes were watering, shaking with tears, which started to drop down her soft, pure skin. "I feel like I don't deserve you, or deserve any happiness. I feel that you shouldn't do stuff that you know might hurt me, but I feel guilty about asking that of you. I feel like I have no idea where to start with my feelings, like every time I try to process them, I run away from them. I feel like I'll always be this way, and I'm trapped, and there's no way out."

I felt my throat collapse. The salty tears trickling down my nose and cheek and chin. We moved our bodies closer together.

Her words pierced my skin like knives, but I waited for her to finish. She wasn't done. I knew she wasn't done. There was something she wasn't saying. I could see it in her eyes. It was evident in the way she looked at me, and I was afraid of the unleashed sentences.

But they didn't come. She didn't say them to me, and that was what broke me.

"I'm sorry," I whimpered. "I never, ever meant to hurt you."

I didn't know what else to say, what else to do. I just reached for her hand and, very slowly, I interlocked my fingers with hers.

I shook my head. I felt guilty, confused, overtaken with sadness. I felt the insect.

Was I passing the insect on to her? Had I done this?

I'd never wanted her to feel this way, and I'd do whatever I could to make sure she didn't anymore.

"I'm sorry, too," she muttered.

We leant towards each other.

"We'll work this out, OK? We will."

33

WATER

Aisling

Edinburgh
AUGUST 2014

I COULD FEEL myself
cracking,
her
slipping away from me.
Further and further
away.
I could feel her.
Ma.
In my brain,
coming for me.
Coming to take me away from Maya.
Or maybe taking
Maya for herself.
Perhaps she was coming for both of us.
I was trying, trying to push her away.
But I was failing.

• • •

BACK IN EDINBURGH, by the time August came around, I felt constantly sick. I loved Maya—I was sure of it. But certain things, they just jumped out at me and brought my past back.

Even the scent of the alcohol brought memories of my mother back to me, which had never happened before. I got nervous when Maya drank.

But it wasn't just that. It was the little things, too. Even something like watching Maya doing the washing up at night, I saw Ma.

I felt her in myself sometimes, too. Her snaps, her anger, her guilt, her jealousy. I saw flashes of her when I looked in the mirror. It made me want to scream, cry, disappear.

Had my mother been right? Would I end up just like her, or worse, with someone who became like her?

I had tried to talk to Maya about how I felt when we'd been in London. I'd felt abandoned by her when she went to see Ethan, especially given that the weekend was for us to spend time together.

When she'd asked how I felt as I lay on her bed, it'd just come out. It felt good to express it, even if I hadn't told her how much my mother was infiltrating my thoughts. I don't think I'd ever been so honest with anyone. The only reason I'd said all that was because she'd been right; I needed to stop blocking her out.

I knew there was stuff bothering her, too. I knew she had been feeling more anxious recently. I could tell from her mannerisms and her increased drinking. I hoped, desperately hoped, that I hadn't caused it or somehow projected it onto her. I wanted to ask her how she was feeling, like she'd done with me, but I was scared about what she'd say. In my most paranoid moments, I wondered if she was avoiding me. Sometimes I felt so scared she might call things off, tell me it wasn't working. When we had fights, she had a habit of storming off. Often, she'd go to the next room, but sometimes she'd even leave the flat altogether to go for a walk.

Perhaps with her life being as perfect as it was, with her supportive parents and her amazing friends, I just wasn't fitting into it any longer.

Since that conversation, not much had changed. It had been a couple of weeks since then, and Maya and I had barely seen each other. Maya was working at the Edinburgh Fringe, in the Assembly venues on George Square. If she was working evenings, she'd often go for drinks with people after her shift, without me. I didn't know any of them really. She'd invited me at first but had stopped after my numerous rejections. Most days, I worked at the bookshop.

Everything felt so unclear and overwhelming. Like logs and kindling were all being loaded up onto a fire, one which was about to be tackled by a flaming phoenix which would set it alight. Whoosh.

Sometimes, though, things would draw me back and make me remember why I wanted to fight for our relationship. It would return momentarily to how things had been before. I would bring her a coffee in bed, and we would talk about things as normal, books and Fringe shows and plans, or she would cook me dinner and light candles and then we'd have sex slowly and quietly. It was in these moments that my emotions ran high; I felt our love for each other, and I wanted to be with her more than anything.

"You know that holiday in the Highlands we were talking about?" she murmured to me as we lay in bed one warm August night.

We were holding hands, but hers had started to slip out of

mine. I clung onto it, dragging it back up onto my chest and curling my fingers around it. The curtains were open, and the window also open a crack, letting in some fresh air.

"Mm-hmm," I said, my eyes fluttering.

"What would you say if I told you I'd booked for us to go this weekend, for your birthday?"

I sat up in bed, suddenly awake.

"What?" I said. "Are you serious, Maya?"

"I am very serious," she laughed.

I squinted at her silhouette in the darkness, the moonlight carving her outline. I could see the faint glow of her smile, and her arms propped up behind her head. I caressed her face with my hands and kissed her, feeling her arms wrap themselves around my body.

"I've asked for a couple of days off and I know you're not working this weekend. Gabe is letting us borrow his car. I've gone on the insurance and rented a little cottage near Glencoe for two nights."

"Thank you, Maya," I whispered. "Thank you. I can't wait."

• • •

THE DAY BEFORE my birthday was the day before we left for the Highlands. Maya was working at the Fringe all day, but she'd said she would be back early evening to pack and get ready to go the following morning.

It was a beautiful summer's day, the day before I turned nineteen.

I looked out the window as I sipped my coffee. The sky was so clear, like a blossoming cornflower. I took another mouthful and thought about the fact that I had the day to myself. I wasn't working at the bookshop again until the following week; I had holiday to use up and had already asked to take these few days off, even before Maya had planned our trip.

I decided I was going to walk to the beach. I could have gone to see some shows at the Fringe, but the day was too beautiful to be inside, I thought.

I got dressed, putting on my swimsuit underneath my clothes; then I packed my backpack. Towel, water bottle, sun cream, sandwich, apple, underpants, and a change of clothes. It'd take me about an hour and a half to walk to Portobello Beach, the map on my phone told me.

I set off, curving around Holyrood Park, hugging the looming hill which towered above me, through Duddingston, and finally into Portobello and onto the beach. It was quite busy on such a sunny day, so I walked along until I found somewhere quieter. I laid my towel down and sat on it, viewing the sea in front of me and fidgeting my feet. I took in the colors. Fluffy waves, fizzing like cream soda, the sand like sawdust, dappled with footsteps.

I must have sat there for an hour or so. Just thinking and watching.

As I stared out at the indigo water, I thought about why I had been keeping so much inside of me. I began to realize that me not talking about things, me not asking about things, was the very quality that made me more like my mother. Perhaps I could beat her, get her out of my head, stop her invading my thoughts, if I just talked and listened more. I could run towards Maya and not away from her. Maybe I could open up to her, or even someone else, maybe a therapist. Maybe I was enough, like Maya had always made me feel. Maybe this would stop me feeling the things my mother felt: anger, jealousy, sadness. Maybe I could love myself and let myself be loved.

But as I sat there, Mother whispered in my ear, telling me I couldn't, instructing me not to share anything with anyone.

I took the water out of my bag and had a long drink. I was thirsty from the walk, and the heat was making me sweaty. Then I reapplied some sun cream to my face. I was starting to burn a bit; I could feel the tenderness on my nose and cheeks. As I opened my bag again to put them back, I saw the apple in there. I picked it out and took a huge bite. I ate the whole thing, savoring each mouthful and then tipping the seeds into my hand from the browning core.

I stood up and let the seeds drop at my feet and looked at them, stripping off my clothes down to my swimsuit.

I ran towards the water.

I didn't run away from it.

I ran

and ran

and ran in.

I kept running even when the water hit me. Even when the waves smacked my legs and stomach. The bitter cold stung me and froze my blood, making my skin red and raw. Then I walked; then I swam. I swam with my head above the water. I could hold my head above the water; I could go under if I wanted and come back up without being held down.

I laughed with joy.

My stomach pumping as breath—

BREATH

—rolled around my body and out of my mouth in little bursts.

I let out small shrieks.

I could breathe.

Then I put my head under. I put my head under the water. I opened my eyes. Into the salt. I couldn't see anything, and they stung like hell, but I brought my head back up and went with the tide back to the shore. As I swam, my eyes started to release wet, salty tears. My body convulsed. I stood, and then strode out of the water, covered less and less by the sea. My eyes still aching, my stomach still rising and falling, I walked back to my towel and rubbed my face with its softness, then wrapped it around me. I stood there, shivering, looking out to the sea and to the horizon beyond.

Part Five

One

34

THE HIGHLANDS

WE DROVE THAT day to the Highlands, around the curving roads and through the lilac, heather-sprung hills. Listening to Nick Drake, we watched the light get caught by feathers of birds and bodies of water. The water had a silver complexion, transformed by bursts of sun into blue dolphin-like ripples.

When we arrived there, we got snug in our wonky cottage, which was tucked within a small town near the loch. The house had blue window frames and beige sheets which smelled like they might have been fresh perhaps a week ago. We clambered underneath them when we arrived and found our way into, through, around each other, molding our bodies. But she felt distant, like smoke I couldn't quite grasp.

Perching herself next to me on the sofa afterwards, she crossed her legs and pulled the tartan rug over our bodies. She took a sip of hot cocoa and our eyes met.

"Everything alright?"

A nod.

"Happy birthday."

Later, the clangs of pans and cutlery.

"Could you help me bring things through?"

"Coming."

The tartan tablecloth being straightened out. A single candle in the middle, dancing amid the shadows.

The smell of alcohol burning in the pan, the smell of soil washing off vegetables.

Turning the heat down on the hob and watching the flame fizzle and flail into nonexistence.

Putting on the oven gloves one by one and picking up the big, bubbling dish, which splattered beneath the lid.

Changing the positions of the knives and forks.

Staring at each other, at the meal.

Quietness hanging in the air.

"Sure everything's OK?"

A nod.

"Aren't you hungry?"

A shake of the head.

That evening passed in near silence. We watched a film and then went to bed. Switched the lights off, our heads facing in opposite directions.

I could feel the distance between us enveloping the entire cottage.

I slept like a child that night, a child being swung from side to side, lilting like a lullaby. I felt exhausted from the confusion, the distance. It had been going on too long. Something had been breaking and it was about to give.

When I woke up the next morning, I rolled over and the bed was half empty. I stared at the space. I stared at it for a long time, absorbing the gap. Absorbing the meaning of the gap.

I knew this was no ordinary gap. This was not a gone-to-the-loo gap, or a making-breakfast gap, or even a making-us-coffee gap. It was the gap of someone who's left.

I looked for her bag, and sure enough, it was gone.

I wished it away, the gap. I thought myself to be imagining it. I rubbed my eyes. I thought, If I fall asleep again, and wake up again, it won't be there. The gap. It will be filled with her, as it should be.

But I got up. I had to get up. I searched. I looked everywhere. My vision hazy with tears which started to drip like blobs of oil down my skin. My brain cloudy, my heart confused, I stumbled and shouted and called out.

"We can get through this," I shouted.

We can. We can do it. I can do it. I can change. I can be better.

I opened the front door, and barefoot, I ran outside. I glimpsed in every direction, my eyes unable to focus.

I just looked for a shape. One shape. Her shape.

I ran down the road. I started to see people, but not her. Cars going past. Days going on as normal.

Eventually, I stopped.

I called her name.

But she was gone.

I called her name.

But she couldn't hear me.

35

WRITING

Edinburgh
APRIL 2015

I WALKED TO meet Gabe for a coffee. Edinburgh had that early-spring sparkle. The brightness of the day was accompanied by a bitterly swift breeze, which made the trees sing a symphony of rustles and whispers.

Gabe had moved in with Isla by that point, who I didn't see much of anymore. I'd stopped going to poetry society. I couldn't face it, not after everything. But Gabe and I saw each other every now and then for a catch-up. I knew he still saw her as well. They had the same ritual as we did: the occasional coffee or walk. But we never really spoke about that, or about her. He'd told us both he didn't want to get involved.

I strolled through Princes Street Gardens. I was meeting him at a café on the other side of town, a place on Hanover Street. The gardens bloomed with a riot of color. The blossoming trees blushed like juicy watermelon flesh, the daffodils dipping their peach and yellow heads down to the sun. I ambled past the shops, wove around the people, and eventually arrived.

Pushing open the door, I saw Gabe on the other side of the

coffee shop. He'd already bought my drink for me, which sat there, pristine, in a tall white mug.

"Hey!"

He looked up and waved at me. "Oh, hello!"

"Thanks," I said, a little embarrassed, pointing to the coffee.

"Oh," he said, flicking his hand. "Don't mention it."

I sat down, took off my long coat, and hung it on the back of my chair. "How's it going?"

Gabe told me about how he and Isla were doing, how he'd just been home to see his mum, and all about his plans for the upcoming summer. Isla and he were going to the South of France on holiday, and then he was traveling down to Cornwall for a bit, where she lived, to meet her parents for the first time.

"They're gonna be big fans of yours. No doubt about it."

He simpered nervously. "I hope so." He smiled. "Oh, and Isla also won this poetry competition the other week, which is exciting."

"That's great! Tell her congrats from me. I can't say I'm surprised, though, she's so talented."

I raised my eyebrows and drank some coffee. I really did miss hearing her writing.

"Isla says they miss you at the meetings."

I looked at him briefly, then quickly peered down at the table. I scratched at a dent in the woodwork with my nail.

"I know, but I can't go back. I just don't think I'm ready to see her yet."

Gabe watched me without moving his head. "I think she really misses you."

I couldn't bring my eyes to meet his. I just let them rest on the dent in the table. I thought about how that dent might have been formed. Had someone dropped something onto the table by mistake? Or done it purposefully, with a knife or something?

I thought in that moment, again, about what had happened last summer. I'd gone over it a thousand times. It kept me awake at night, trying to figure it out. Trying, desperately, to understand what had happened and why she'd left. Maybe she felt I was becoming someone she didn't want to be with, or perhaps she sensed herself becoming that person, or it even could've been that she thought I was convinced she was becoming that person. I was still confused, and I likely always would be. Maybe, with these things, you never feel as if there's a good explanation.

"Why, did she say that to you, or—that she misses me, I mean?"

I paused. Then I did look up at him. Gabe nodded. His eyes seemed soft and gentle, magnified through his glasses.

"I know I said I won't get involved," he mumbled, sitting back in his chair, "but I thought you should know."

"I miss her, too. But I'm not ready to—"

"I know," he said. "I get that."

After a moment of hush, Gabe asked me a question: "How do you normally process things, understand things?"

I sat back in my chair and gazed at him.

There was one constant.

After burying so much for so long, I had realized over the last year that one thing, especially, allowed me to understand. One thing enabled me to process what happened to me and what happened around me. It was through this thing that I could attempt to comprehend the reasons why people did or said things. It empowered me to see life in more vibrant colors, to amplify sounds, to relive sensations.

"Writing, I guess."

Gabe nodded. "Why don't you write it all down then?"

"What, like a book?" I laughed.

Gabe didn't laugh back. "Yeah. Just, something. Something about her."

He paused, smiled, and eventually shrugged. "Only if it might help you to understand what happened, though," he whispered.

As Gabe smiled softly at me, I thought about what he'd just suggested. I let his words sit there in the air around us. I let them twist and turn like ballerinas. I let them swirl and spiral, like mist which hovers over the grass in the early morning.

Perhaps, I thought. Perhaps I can try.

ACKNOWLEDGMENTS

There was a time I was certain that this book would never be published. That time has made it all the more special having *Something About Her* out in the world, and it has made me feel so profoundly grateful to the people who have got it here. I really hope I can put my heartfelt gratitude into words.

I could not have wished for a more incredible, dedicated agent than Millie Hoskins. Thank you so much, Millie. The way in which you have worked with me and believed in me is extremely special. I am so grateful for your support and guidance, and for absolutely everything you have put into this book.

Thank you also to Becky Percival at United Agents, who gave such discerning and insightful comments on the original manuscript, including pointing me towards the enchanting work of Katherine Mansfield, which became the epigraph.

I also want to express my deepest gratitude to Caradoc King. Your kind words made me believe I had an important story to tell. Thank you for believing in me and in this book from the very beginning.

To my wonderful editor, Gabriella Mongelli. Gaby, you have patiently and skillfully transformed this book into more than I could ever have hoped for. Thank you for putting so much into every single line of text and for instantly understanding the people in these pages as soon as you met them. There is no one with whom I'd rather have shared this journey. Working with you has been a joy. Thank you, thank you, a million times over.

To the brilliant team at Putnam: Sally Kim, Ashley McClay, Nicole Biton, Alexis Welby, Brennin Cummings, Emily Leopold, Marie Finamore, Alison Cnockaert, Christopher Lin, Mary Beth Constant, and LeeAnn Pemberton. Thank you all—I am so grateful for the magic you have worked on this book and for all your hard work.

Writing this book was not something I shared with many friends, but there were a few people who knew, and without whom I don't think I would have had the courage to keep going.

Thank you, Jenny Quested and Laura Milford, for reading early chapter drafts, for not laughing in my face, and for being so kind with your comments and encouragement. I am very lucky to have your friendship.

Johnny Staunton Sykes, you've been there every step of the way. You believed I was capable of writing this book a long time before I did. Thank you for everything. I am certain that this novel would not exist without you.

My deepest thanks also go to Alice Peet, Belinda Quested, Fei Yen Waller, Claudia Gleeson, Orla Woodward, Jake Moscrop, and James Costello O'Reilly, some of whom were unfortunate enough to live with me while I was writing. I hope you all know how much you mean to me. Thank you for the cups of tea, the cake, the discussions, the poems, the music, the walks, the cycles, the laughter, and generally some of the happiest times of my life.

Thank you to Annie Shipton, for not only being a fantastic

friend but for answering all my many questions about being an undergraduate student at Edinburgh University. Thank you also to all those whom I have shared time with in Edinburgh over the years, especially to those who have kindly welcomed me to stay (special thanks to Peter, Murat, and Susan for their generosity). Edinburgh is a city which continues to bewitch me. I hope the book begins to do justice to its magic and beauty.

My gratitude also to Gemma Turnbull and Matyas Molnar. Thank you both for being so generous with your time and skill.

I am grateful to my doctoral supervisors, past and present, Professor Gina Neff and Professor Vicki Nash. Thank you for being such supportive and inspiring mentors, and for encouraging me to swim in my own lane.

Last being the opposite of least, thank you above all to my family—Peter, Jill, Katie. You mean everything to me. Your kindness, selfnessness, brilliance, and joie de vivre astound me every single day. There's no doubt about it: I couldn't and wouldn't have done any of this without being inspired by your extraordinary minds and hearts. Thank you especially to my parents, who have given and given up so much for my sister and me over the years.

To all my friends and loved ones, including those named above and those not, thank you for being there and for being exactly as you are. I'd like to dedicate this book to your friendship, and to your strength and resilience, which I've witnessed firsthand when life's complexities have been thrown at you. I've encountered so many courageous people in my life. I know there are also plenty out there whom I've never met as well, and this book is dedicated to you, too.

I know this might sound crazy, but I have to say it. Thank you, Aisling and Maya. I know you are characters in a book to most, but you are real to me. Spending time with you has been one of the greatest pleasures of my life so far.